SHAKE UP

SCIENCE 1

Pearson Education Limited
Edinburgh Gate
Harlow
Essex CM20 2JE
England
and Associated Companies throughout the world.

www.pearsonelt.com

First published 2016
ISBN: 978-1-2921-4470-2
Set in Futura LT Pro, Feltpen Com, Bauhaus Std, ITC Benguiat Gothic Std, ITC Zapf Dingbats Std

Acknowledgments
Picture Credits

The publisher would like to thank the following for their kind permission to reproduce their photographs:

(Key: b-bottom; c-center; l-left; r-right; t-top)

Student's Book: 123RF.com: 6tl, 10bl, 11tr, 13bl, 17t, 17b, 18tl, 27cl, 28tl, 28cl, 28bl, 30tc, 30cr, 31tr, 33c (left), 33cr, 33bc, 34l, 34c, 34r, 39 (cat), 39 (sandwich), 40tr, 40bl, 48c, 51tl, 51bl, 53l, 54b, 55tr, 55br, 57tl, 57tr, 58t, 59br, 60l, 60b, 61cr, 63c, 64tl, 64c (left), 64cl, 65tc (left), 65tc (right), 65cl, 67tl, 68l, 71tl, 75tl, 77tl, 77tc, 77tr, 77bl, 78tl, 78bl, 79bl, 79br, 80tl (right), 80bl (right), 80bc (left), 80bc (right), 81tl, 81tc, 81tr, 81bc, 82tc, 82bl, 83t, 83br, 84tl, 84c, 84bl, 85tr, 88tc, 88bl, 89t, 90t, 92t, 92cl, 96cr (bottom), 99bl, 99br, 107tr, 108br, 109c (bottom), alekleks 97cr, Scott Betts 55tl, 63t, Sergiy Bykhunenko 53c, fckncg 100c (right), Mike Flippo 100c, ftlaudgirl 104cr, Eric Isselee 66bc, jirkaejc 58r, Vladimir Kramin 100c (left), Vera Kuttelvaserova Stuchelova 102t, Pavel Losevsky 53cl, Lucian Milasan 64cr, Janunya Napapong 109tc, nito500 62cl, Rafał Olechowski 53r, Thomas Pajot 87br, Dario Lo Presti 56tr, Harris Shiffman 52t, Ljupco Smokovski 105/1, Stanimir Stoev 7c (bottom), 13tl, Maksym Topchii 53c; **Alamy Images:** TongRo Image Stock 106br; **Fotolia.com:** anake 67b, Steve Byland, Jacek Chabraszewski 103b, dearzkung 76t, dennisjacobsen 29br, donvanstaden 29bl, famveldman 29t, gilmart 94cl, Gray Wall Studio 67tc, 71tc (right), iofoto 6tc, jolopes 39tl, katyspichal 105tc, Ivan Kmit 35bl, 37cl, Konstiantyn 101bl, Brian Kuns 28br, Konstantinos Moraiti 31b, neirfy, Sergey Novikov 111b, Popova Olga 96c, Paul Prescott 105tr, 107bl, pressmaster 25c, PRILL Medienmedesign 108tc, 109bc, ra3rn 93tc, Sasajo 80tc (right), scabrn 42t, snaptitude 61tl, Swapan 80tr (right), Szasz-Fabian Jozsef 105/2, V&P Photo Studio 25l, 27tr, WavebreakmediaMicro 27br, 30cl, 39cr, winston 91br, Vladimir Wrangel 18tr, yvdavid 91tl, Tetiana Zbrodko 35cl, 36tr, Mara Zemgaliete 28cr, 33cl; **Imagemore Co., Ltd:** 4tc, 9tc (left), 23bl, 33tr, 89r; **Imagestate Media:** John Foxx Collection 5bc, 69l, 81cl, 107cl; **MIXA Co., Ltd:** 72tl; **Pearson Education Ltd:** Studio 8 76b, Pearson Education, Inc. 91bl, Sian Bradfield. Pearson Education Australia Pty Ltd 27tl, Tudor Photography 85c, 89br, Jules Selmes 59tr, Rafal Trubisz 105b, Jesus Ismael Vazquez Sanchez 24r; **SF Glenview Photo Studio:** 14, 26, 50, 74, 86, 98, 110b; **Shutterstock.com:** 3445128471 9b, 13tr, 4Max 78 (spoon), Africa Studio 4c (left), 52 (ear), AJE 93br, aldegonde 56bl, 80tl (left), Ermolaev Alexander 81bl, AlexAvich 85bl, Hintau Aliaksei 7cl (top), 66br, AM-STUDiO 52bc (left), 61cl, Volnukhin Anatoly 37cr, Andresr 42c, Andy Dean Photography 106tr, Andy-pix 105/3, Venus Angel 80br (left), 82tr, arek_malang 56cr, 77br, arka38 28bc, 35br, 40c, Noam Armonn 99bc, 107c (left), artjazz 58br, aza_za 94tc, B747 71br, Paul Banton 54bc, Joe Belanger 56tl, Henk Bentlage 61bl, Stephane Bidouze 7cr, bikeriderlondon 92c, Ruth Black 80tr (left), Bloom Design 87bl, BlueOrange Studio 73b, Ekaterina V. Borisova 65c, BortN66 106cl, Steve Byland 31tl, 38cr, Rob Byron 52bl, 57cr, Cameramannz 41bl, Rich Carey 5bl, carroteater 16tl, Marek Cech 95tr, Diego Cervo 105c, Sam Chadwick 106c, Hung Chung Chih 30b, clearimages 58c, Cococinema 85tl, Matthew Cole 62, 78cl, 100bl, 108t, 108bl, 109t, Coprid

52bc (right), 78 (jug), 88tl, 94c, Paul Cotney 7bc, 13cl, Creative Travel Projects 51br, David Crockett 56cl, cynoclub 57br, 65bc, danielo 64c (right), Francesco De Marco 85br, Michiel de Wit 44cr, 45cl, DenisDV 56br, Dinga 88cl, Le Do 6cr (top), dotshock 76l, Denis Dryashkin 109c (top), Jaimie Duplass 42b, Lane V. Erickson 42cr, Richard Evans 48l, 51tr, David Evison 80br (right), 82br, Melinda Fawver 44tr, 55bc, 87bc (left), Germany Feng 87t, George Filyagin 97b, 110c, Bill Florence 36bc, Svetlana Foote 97tl, fotohunter 47b, Fotokostic 100tl, 100tr, 100br, 101cr, G. K. 70t, Gelpi JM 36br, Deyan Georgiev 92b, Gladskikh Tatiana 78cr, godrick 67tr, Warren Goldswain 104cl, gorillaimages 92cr, Andrii Gorulko 82bc, Joe Gough 58bl, Grebnev 27bc, Guzel Studio 101t, Gyvafoto 57bc, HABRDA 70bl, Harper 3D 62br, Mat Hayward 19tr, 60cl, Ramona Heim 7tl, Brent Hofacker 22cl (bottom), Jiang Hongyan 9tr, 22br, 57bl, Anna Hoychuk 88bc (left), Hurst Photo 18br, 24l, Rob Hyrons 23tr, Peter Igel 72bl, Aleksandra Imanutinova 23tc, iofoto 106tl, Ionia 96cr (top), ISchmidt 104tr, Eric Isselee 7cl (bottom), 8bl, 32l, 32r, 40tl, 40cl, 40cr, 41br, 44tl, 45r, 55bl, 65bl, 80bl (left), Jack.Q 89c, Boumen Japet 16br, johnfoto18 96bl, joppo 83bl, Szasz-Fabian Jozsef 64br, jps 10br, 41tr, Kachalkina Veronika 105/4, Alex Kalashnikov 83bc, Stepan Kapl 57cl, kavram 68c, kavring 84tr, Brian Kinney 100cr, 107cr, Kjuuurs 6br, Alex Emanuel Koch 37tr, Ruslan Kokarev 69r, Konstanttin 72tr, koosen 39 (water), 94tr, Eduard Kyslynskyy, Kzenon 60cr, Laborant 88tr, Erik Lam 56bc, Lucie Lang 22bl, Katrina Leigh 94b, Lindasj22 39 (house), Jim Lopes 70br, magicoven 23tl, 81c, marcioalves 76c, mariait 28tr, Rob Marmion 11bl, Jan Martin Will 38bl, Mircea Maties 108bc, 109b, Marco Mayer 57tc (left), Mazzzur 40bc, 48r, Mega Pixel 4tl, 9tl, Aleksandar Mijatovic 4tr, Stuart Miles 87bc (right), mimo 63b, MNStudio 71bl, Steve Mollin 95bl, 99tr, Monkey Business Images 27bl, 106bl, Morgan Lane Photography 61br, motorolka 96br, Andrii Muzyka 65tl, Natalia61 94cr, Natursports 10t, 11tc, NIK 100cl, nikshor 22bc, Nordling 78tr, Martin Novak 40br, Olga_Phoenix 58bc, Tyler Olson 5tr, 107tc, Chepurnova Oxana 18bc, paffy 4br, 84br, Damian Palus 41tl, 66bl, panbazil 103t, 111t, Ildi Papp 8cr, 9tc (right), Anita Patterson Peppers 7tr, paulaphoto 60r, Pavel L Photo and Video 73t, Siamionau Pavel 52cr, 96tr, pbombaert 4cl, 52 (eye), pedrosala 54tc, M. Pellinni 64bl, 68r, Catalin Petolea 61tc, Dasha Petrenko 54t, 75br, photomaster 43bc, 43br, photosync 18bl, Picsfive 38tl (top), 43bl, 50tl, 110t, picturepartners 102b, 106bc, 107br, Pikoso.kz 64tr, 66tr, 75tr, Olga Polyakova 61tr, Leigh Prather 65br, Pressmaster 7br, 15t, 16bl, 79tc, Procy 69c, Inara Prusakova 90r, 99tl, Denys Prykhodov 4c (right), 52 (hand), Zeljko Radojko 39b, 47t, reptiles4all 6cr (bottom), 33br, Fesus Robert 39tr, Rossario 93b, Federico Rostagno 79tr, s-ts 88bc (right), Artem Samokhvalov 11br, Ed Samuel 88br, schankz 4c, 4cr, 52 (nose), 52 (tongue), Ken Schulze 101br, Daniel Schweinert 81cr, Ilin Sergey 13b, Saied Shahin Kiya 43tr, shnjr52 79tl, Roman Sigaev 95cr, Dr Ajay Kumar Singh 4bl, KC Slagle 99cr, somchaij 80tc (left), 82tl, SOMMAI 8tr, Alex Staroseltsev 91tr, studioVin 22tl, 38tl (bottom), 108cr, Lisa A. Svara 38cl, Bayanova Svetlana 35cr, Taiga 6tr, tanuha2001 44br, Tatiana53 36bl, 37bl, tentor 57tc (right), Thaiview 81br, Anatoly Tiplyashin 36cr, 38br, Tomsickova Tatyana 104bl, topseller 96cl, tratong 89bl, Triff 57c, 71tc (left), Suzanne Tucker 7tc, Lisa Turay 52br, Pamela Uyttendaele 23bc, Valio 71tr, lakobchuk Viacheslav 20, Dmytro Vietrov 18tc, VikaRayu 28cr, vilax 35t, visionaryft 59bc, VR Photos 7bl, 11tl, 33bl, warayut 65tr, wavebreakmedia 5br, xpixel 4bc, 4bc (left), 7c (top), 8cl, 8br, 32c, 33c (right), 44c, 45l, Pan Xunbin 44l, 45cr, Wong Sze Yuen 42cl, Serg Zastavkin 72br, Zerbor 37c, Zurijeta 12, 15b, Zush 19bl, Zvyagintsev Sergey 22cl (top); **Sozaijiten:** 96bc, 107c (right); **www.imagesource.com:** 25r, 29c, 59bl

Flash Cards: alekleks, Sergiy Bykhunenko, Pavel Losevsky, Rafał Olechowski, Maksym Topchii; **Fotolia.com:** forcdan, Paul Prescott, winston; **Imagestate Media:** John Foxx Collection; **Pearson Education Ltd:** Lord and Leverett; **Shutterstock.com:** 3445128471, Africa Studio, Noam Armonn, auremar, Sascha Burkard, CandyBox Images, Eldad Carin, Jacek Chabraszewski, Cococinema, Matthew Cole, Peter Dankov, djgis, ekawatchaow, Svetlana Foote, fotohunter, Fotokostic, Frontpage, G. K., Gladskikh Tatiana, Guzel Studio, HABRDA, Mat Hayward, Jiang Hongyan, Rob Hyrons, Aleksandra Imanutinova, ISchmidt, Eric Isselee, Jack.Q, Rene Jansa, joppo, Cathy Keifer, Kzenon, Laborant, leolintang, Jim Lopes, magicoven, marcioalves, Edcel Mayo, nexus 7, nikshor, Tyler Olson, Chepurnova Oxana, pbombaert, Catalin Petolea, photomaster, picturepartners, Olga Polyakova, Potapov Alexander, Leigh Prather, Pressmaster, Procy, Inara Prusakova, Denys Prykhodov, Samuel Borges Photography, schankz, Saied Shahin Kiya, Shutterschock, Roman Sigaev, somchaij, StevenRussellSmithPhotos, Sam Strickler, Vilor, Bogdan Wankowicz, Zush

All other images © Pearson Education

Every effort has been made to trace the copyright holders, and we apologize in advance for any unintentional omissions. We would be pleased to insert the appropriate acknowledgment in any subsequent edition of this publication.

Science Consultant
Mark Sander

Illustrated by
Ismael Vazquez **19**c, **21**c; Marcela Gómez Ruenes Worksheets (Unit 8 origami frog, snowflake)

Cover images: *Front:* Shutterstock.com: Eduard Kyslynskyy l, Valua Vitaly r; *Back:* Fotolia.com: Steve Byland r, neirfy l; **Shutterstock.com:** Taiga c

Contents

Unit 1

The Nature of Science **T2**

What is science?

Let's Investigate! Lab • How do things look? **T14**

Unit 2

Solve Problems **T15e**

How can you solve problems?

Let's Investigate! Lab • How can you lift heavy things? **T26**

Unit 3

Living and Nonliving Things **T27e**

What can you say about living things?

Let's Investigate! Lab • How are animal and plants different? **T38**

Unit 4

Plants and Animals **T39e**

How do living things change as they grow?

Let's Investigate! Lab • How does a butterfly change? **T50**

Unit 5

Body and Senses **T51e**

What am I like?

Let's Investigate! Lab • How many points can you feel? **T62**

Unit 6

Earth and Sky **T63e**

What are Earth and the sky like?

Let's Investigate! Lab • What do the day and night skies look like **T74**

Unit 7

Objects **T75e**

What are objects like?

Let's Investigate! Lab • Which object is heavier? **T86**

Unit 8

Matter and Mixtures **T87e**

What are matter and mixtures?

Let's Investigate! Lab • What is in a mixture? **T98**

Unit 9

Motion **T99e**

What are position and motion?

Let's Investigate! Lab • How can you move the car? **T110**

CONTENT AND LANGUAGE INTEGRATED LEARNING (CLIL)

Increasingly, students around the world who don't speak English at home are learning content subjects such as science through the medium of English, meaning that English language learning is taking place at the same time as the learning of content.

Benefits include:

- exposure to and acquisition of English language in context, encouraging a more natural language learning process
- meaningful use of the English language, with students motivated to use English to find out more about real-world topics that interest them
- increased English fluency through using the language for a variety of purposes and in a number of different ways
- faster and higher-level development of skillswork, especially reading and writing
- preparation for future studies and the international workplace.

Varied support for English language learners is provided throughout the teaching notes, including additional background information, suggestions for suitable language-learning activities, as well as strategies and techniques for developing skillswork.

USING THE MATERIALS

Teaching and learning situations can differ widely, and, with this in mind, the series has been devised to allow teachers the flexibility to customize according to their requirements. Following a modular approach, each lesson can work as a self-standing unit of content, and teachers can pick and choose to fulfill their own curriculums. Fast-track routes can be followed in situations where less time is allocated for the teaching of primary science through English. More information about fast-track routes can be found online.

In addition to the wide range of reinforcement and extension activities provided through the ActiveTeach, an optional **Workbook** is also available. The **Workbook** has been especially tailored for the requirements of English language learners and provides:

- activities relating to each lesson's key vocabulary and concepts
- targeted practice of already known grammar
- comprehensive development of science-related reading and writing skills
- a progression through receptive understanding to productive ability
- an emphasis on real-world application and students' own experience.

Series Components

Student's Book

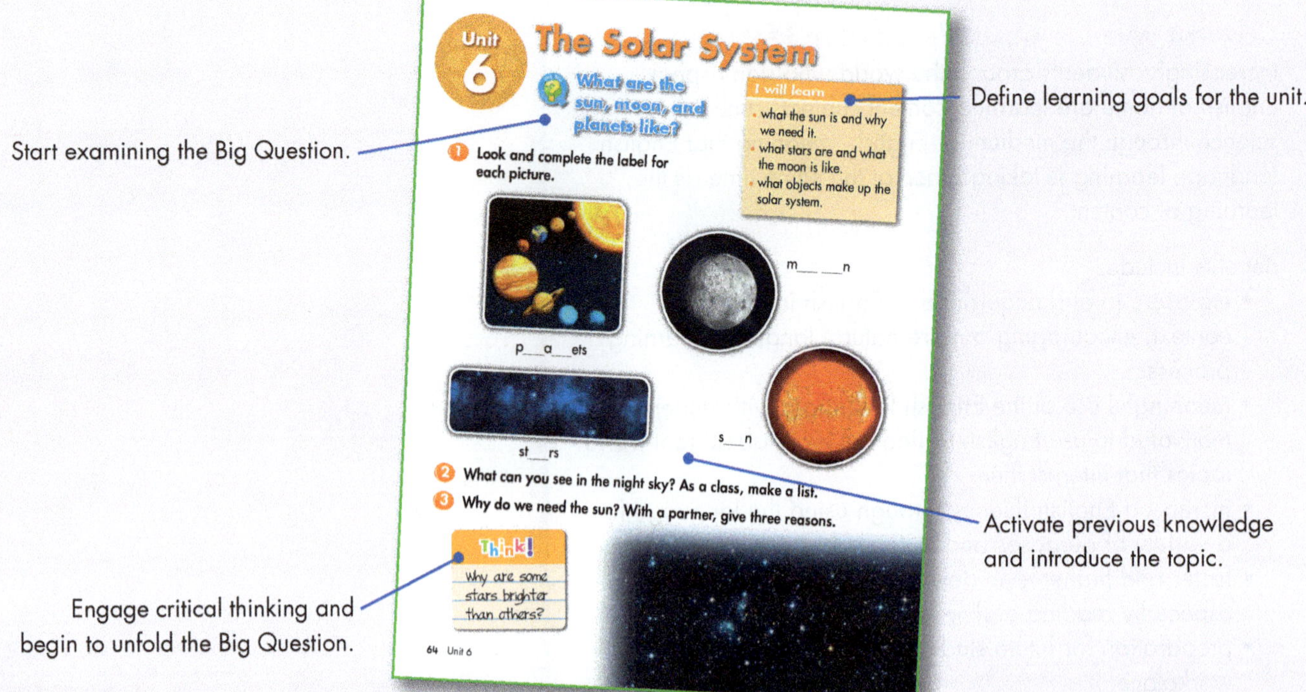

Start examining the Big Question.

Define learning goals for the unit.

Activate previous knowledge and introduce the topic.

Engage critical thinking and begin to unfold the Big Question.

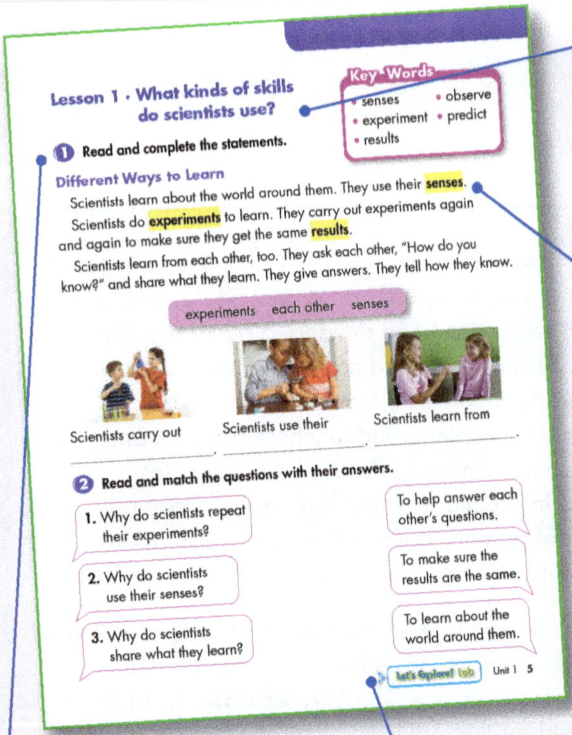

Bring science to life with clearly defined questions, real-world contents, and scientific facts.

Learn key words through texts and definitions in a glossary at the end of each Student's Book.

Experiment in class or online; recording observations in Student's Books gives a sense of ownership. Expand thinking with *Activity Cards* on the ActiveTeach.

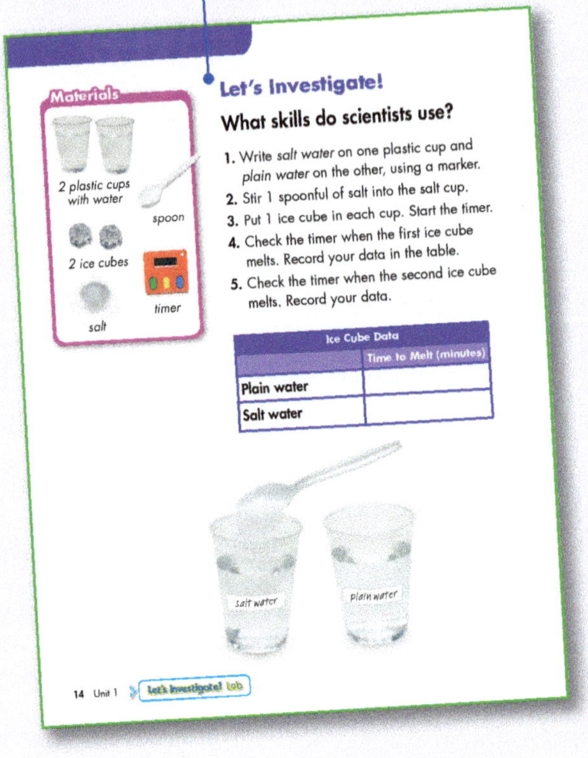

Think, read, and write like a scientist to make learning personal, relevant, and engaging. Explicit instruction brings science concepts to life.

Link to digital activities to explore topics before reading.

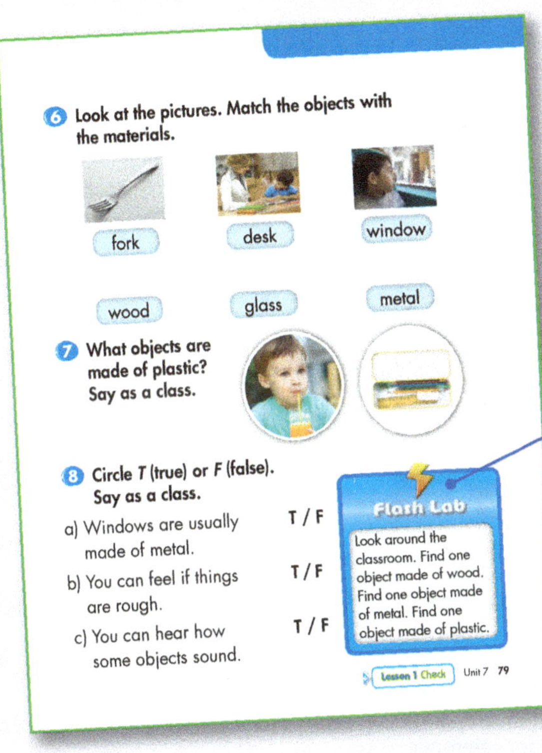

Do quick activities in the classroom.

Introduce or review lesson concepts.

Provoke thought about how to protect Earth.

Do fun experiments with the family.

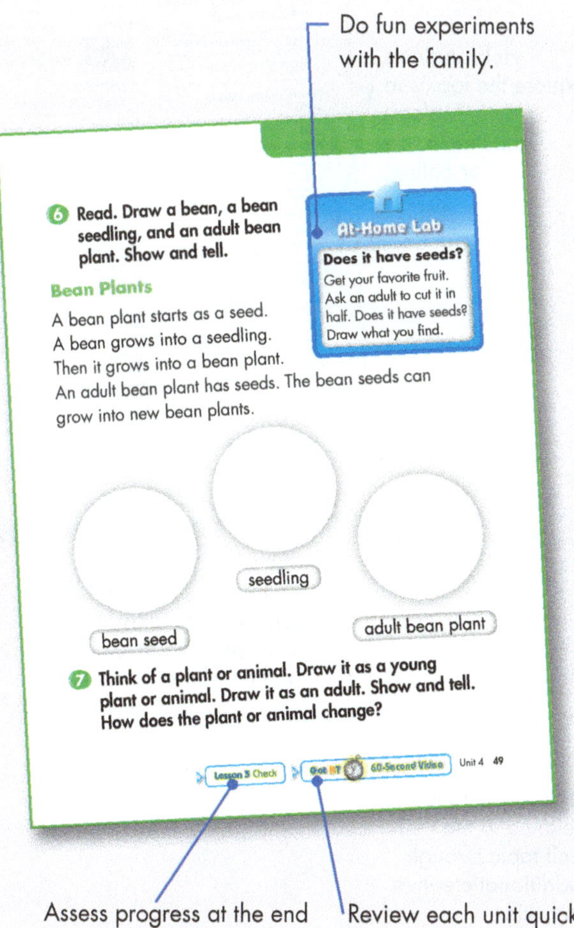

Review the main points of each lesson before taking the Got it? Quiz.

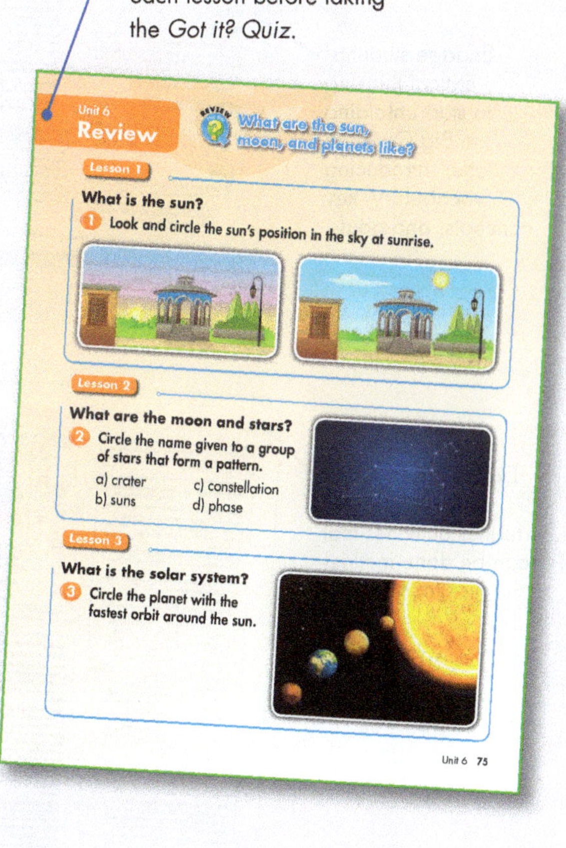

Assess progress at the end of each lesson.

Review each unit quickly and concisely.

Series Components

Teacher's Book

Follow the 5-E methodology (pages xii–xiii) across each level's activities.

Plan your lessons by selecting the activities that best suit your classroom needs, with an estimated time for each activity.

Select Flash Cards for use during the lessons.

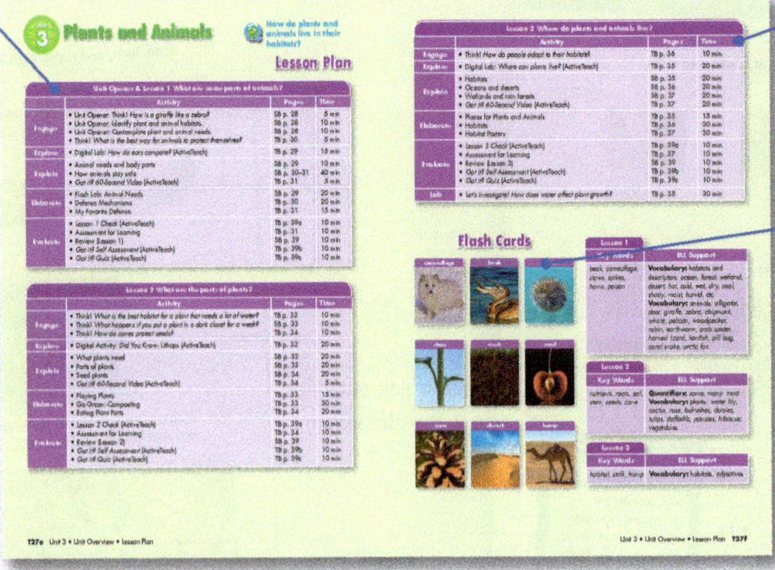

Refer to objectives, vocabulary, and learning resources.

View the annotated Student's Book page for reference.

Engage students' critical thinking to start unfolding the Big Question after introducing vocabulary, key concepts, and goals.

Help students explore the topics in *Let's Explore!* Labs hands-on or online.

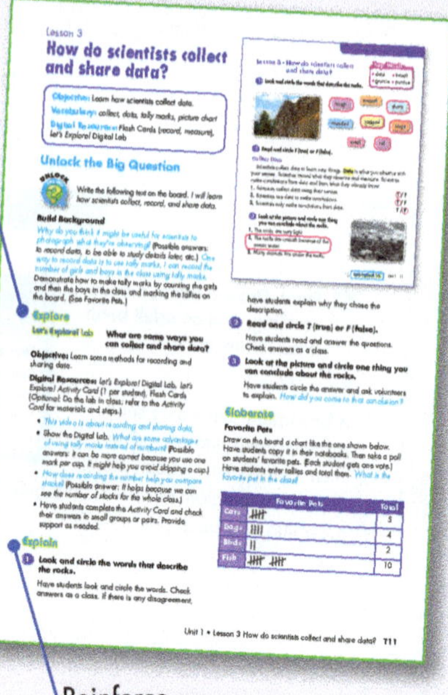

Reinforce understanding of the scientific concepts through core content and activities.

Extend the scientific concepts through further activities that can be done in class or virtually.

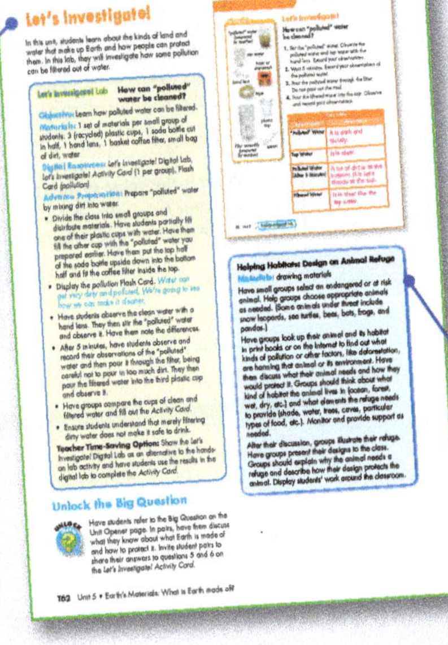

Explore more ideas relating to the unit topic through additional creative activities.

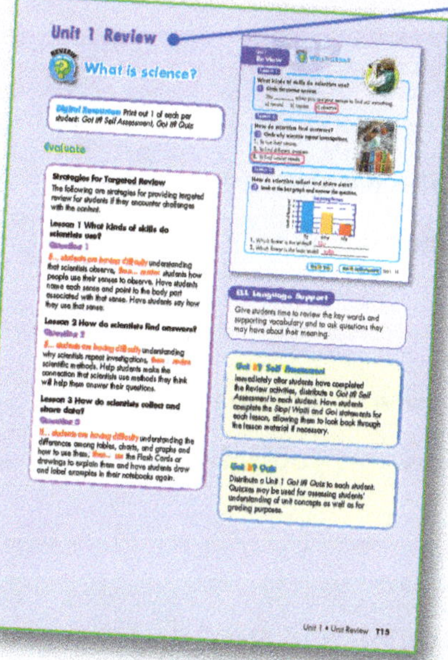

Address challenges students may have while reviewing the unit material and link to *Got it? Self Assessments* and *Got it? Quizzes*.

Deepen students' knowledge and encourage students to elaborate on topics in creative ways.

Link to *Lesson Checks*, *Got it? 60-Second Videos*, as well as Assessment for Learning activities.

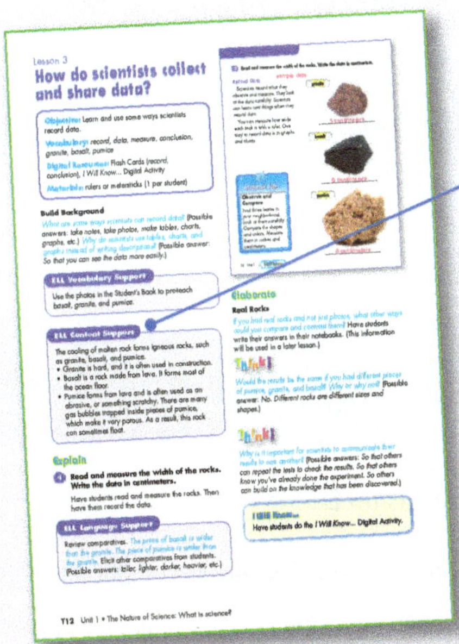
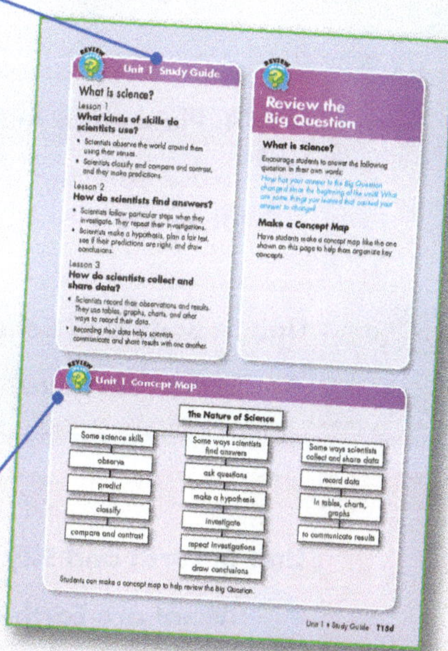

Use Study Guides to summarize the main points in each lesson and review the Big Question.

Support English Language Learners through background information, suitable activities, and skillswork strategies.

Review main unit concepts using concept maps downloaded from the ActiveTeach.

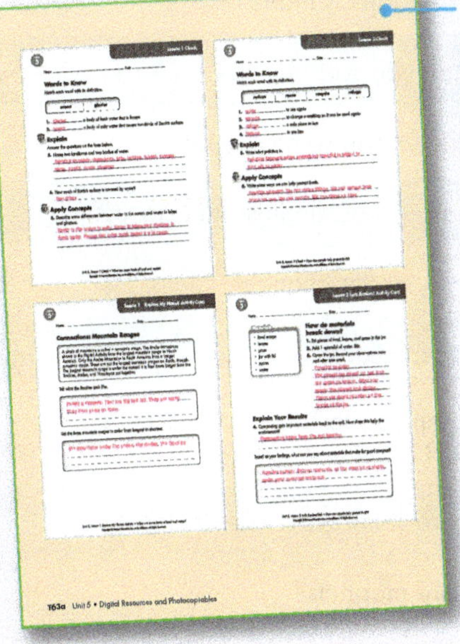
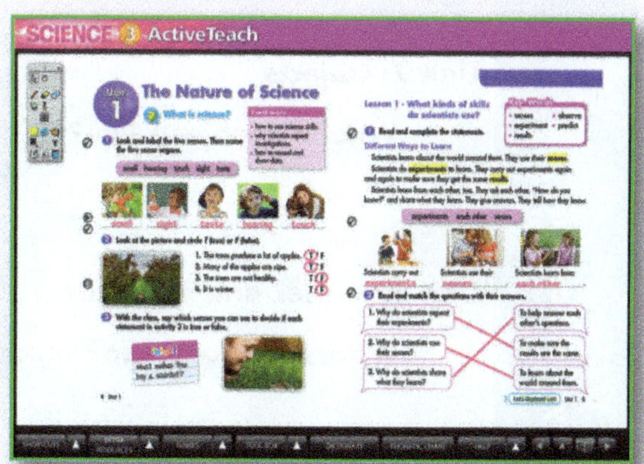

Check understanding and do exploratory activities using cards downloaded from the ActiveTeach.

Access the digital activities, Flash Cards, and all printable resources, including *Activity Cards* and *Quizzes*, in the ActiveTeach.

Scope and Sequence

Units	Lessons
Science, Engineering, and Technology	
Unit 1: The Nature of Science **What is science?**	Lesson 1: What questions do scientists ask?
	Lesson 2: How do scientists observe?
	Lesson 3: How do scientists collect and record data?
Unit 2: Solve Problems **How can you solve problems?**	Lesson 1: What are problems and solutions?
	Lesson 2: How do ideas become solutions?
	Lesson 3: How can you test and share solutions?
Life Science	
Unit 3: Living and Nonliving Things **What can you say about living things?**	Lesson 1: What are living and nonliving things?
	Lesson 2: How are animals alike and different?
	Lesson 3: How are plants alike and different?
Unit 4: Plants and Animals **How do living things change as they grow?**	Lesson 1: Do all young animals look like their parents?
	Lesson 2: How do some animals grow and change?
	Lesson 3: How do some plants grow and change?
Unit 5: Body and Senses **What am I like?**	Lesson 1: What are my senses?
	Lesson 2: What does my body need?
Earth Science	
Unit 6: Earth and Sky **What are Earth and the sky like?**	Lesson 1: What makes up Earth?
	Lesson 2: What can you see in the day and night skies?
	Lesson 3: What is the weather? What are the seasons?
Physical Science	
Unit 7: Objects **What are objects like?**	Lesson 1: What are objects made of?
	Lesson 2: How can you sort objects?
	Lesson 3: How do we use some objects?
Unit 8: Matter and Mixtures **What are matter and mixtures?**	Lesson 1: What are solids, liquids, and gases?
	Lesson 2: How can matter change?
	Lesson 3: What is a mixture?
Unit 9: Motion **What are position and motion**	Lesson 1: What can you tell about an object's position?
	Lesson 2: What are some ways objects move?
	Lesson 3: What are magnets?

I will learn...	Key Words
• that scientists ask questions to learn.	• *scientist, science, observe, objects, questions, answers*
• ways scientists observe things.	• *senses, tools, measure, compare, group*
• ways scientists collect and record data.	• *collect, data, record, chart*
• about problems and solutions.	• *problem, solve, solution*
• how an idea becomes a solution.	• *idea, plan, design, choose, material*
• how to test and share solutions.	• *test, change, share, use*
• about living and nonliving things.	• *living, grow, need, nonliving, move*
• what living things need. • how animals are alike and different.	• *fur, body coverings, feathers, paws, fins, wings, beaks*
• how plants are alike and different.	• *stems, leaves, roots, flowers, seeds, petals, trunks*
• that living things grow and change.	• *look like, parents, butterfly, caterpillar, hatch*
• how some animals grow and change.	• *adult, lay eggs, life cycle, frog, tadpole, chrysalis*
• how some plants grow and change.	• *seedling, fruit*
• about the five senses.	• *see, hear, touch, taste, smell, feel, skin, tongue*
• what my body needs.	• *healthy, energy, exercise, sleep, shelter*
• about kinds of land and water on Earth.	• *Earth, land, oceans, lakes, rivers, swamps*
• what I can see in the day and night skies.	• *sky, sun, clouds, moon, stars*
• about weather and seasons.	• *weather, sunny, cloudy, clear, rainy, windy, snowy, seasons*
• what some objects are like.	• *weigh, heavy, light, wood, plastic, metal, glass*
• how to group some objects.	• *sort*
• some ways to use objects.	• *round, square, strong, see through, clay, sticky, wool*
• about solids, liquids, and gases.	• *matter, solid, liquid, container, gas*
• how water can change.	• *freeze, ice, melt, boil*
• about mixtures.	• *mixture*
• about position.	• *position, on, above, below, in front of, behind, next to*
• about how objects move.	• *push, away from, toward, pull, fast, slow*
• about magnets.	• *magnet, attract, repel*

5E-Methodology

Shake Up Science is based on the 5E-Methodology:
Engage, Explore, Explain, Elaborate, Evaluate.

Engage

On the first page of every unit, students are introduced to
the Big Question, the question that will guide their learning
throughout the unit. On this page, students are encouraged
to start engaging with the topic. *Think!* questions in the
Student's Book and additional questions in the Teacher's
Book help students to think critically and unfold the Big
Question.

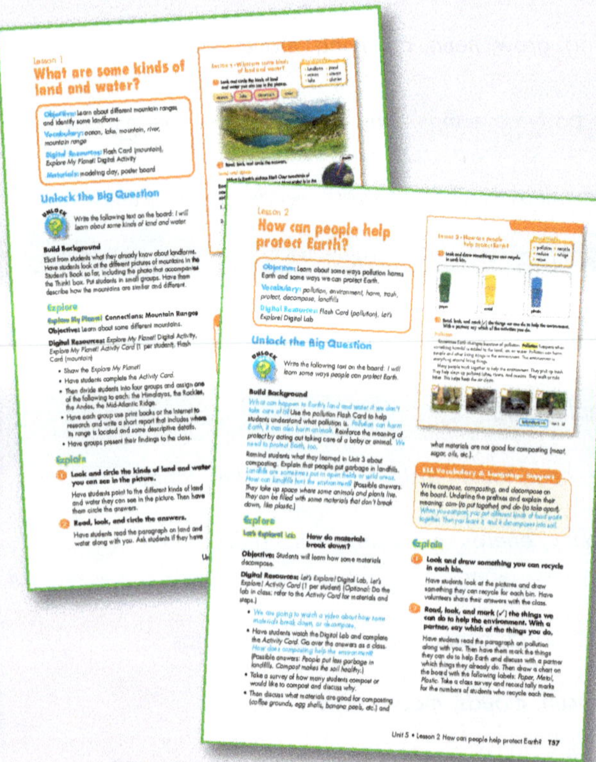

Explore

In every unit, students have an opportunity to explore some
of the main concepts before they start reading core content.
In each unit, there is a *Let's Explore!* Labs that offers students
an opportunity to explore, hands-on, an idea relevant to
the related lesson. Teachers may opt to do the activities in
class or show them via the ActiveTeach. Related *Activity
Cards* that can be printed from the ActiveTeach help further
reinforce students' activity-based learning.

Explain

To build their understanding, students read a variety of texts
that explain scientific concepts. Core content is accompanied
by activities for students to do—individually, in pairs, in small
groups, or as a class—to enhance understanding of the concepts.
ELL Content Support boxes provide background information or
activities teachers may wish to include in their lesson plans. *I
Will Know…* Digital Activities can be used in class to reinforce
and practice the main ideas. *Got it? 60-Second Videos* provide a
comprehensive review of the scientific concepts covered in each
unit. At the end of every unit, *Let's Investigate!* Labs consolidate
unit concepts through hands-on experiments. Teachers can opt
to do the Labs hands-on in class or to show the digital materials
instead. Students reflect on their learning by answering questions
in their Student's Books. Related *Activity Cards* that can be printed
from the ActiveTeach help further extend students' reflection.

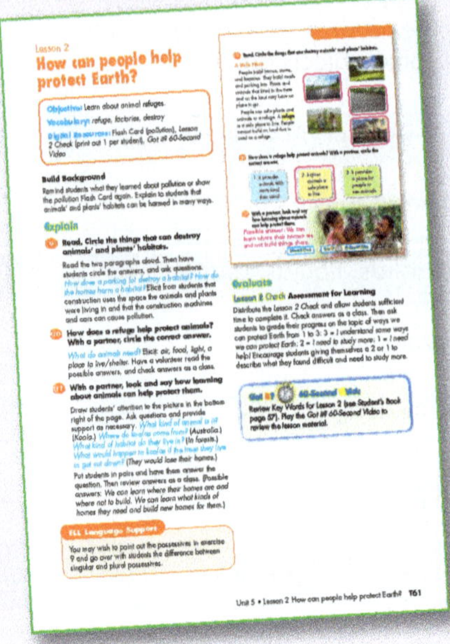

ELABORATE

Throughout the units are a number of activities designed to deepen students' knowledge of the topic in fun and creative ways. In addition, class projects and other kinds of creative activities are offered at the end of every unit. Students are given various opportunities to write, perform skits, make murals, give presentations, do research, compose poems, and design inventions, among other things.

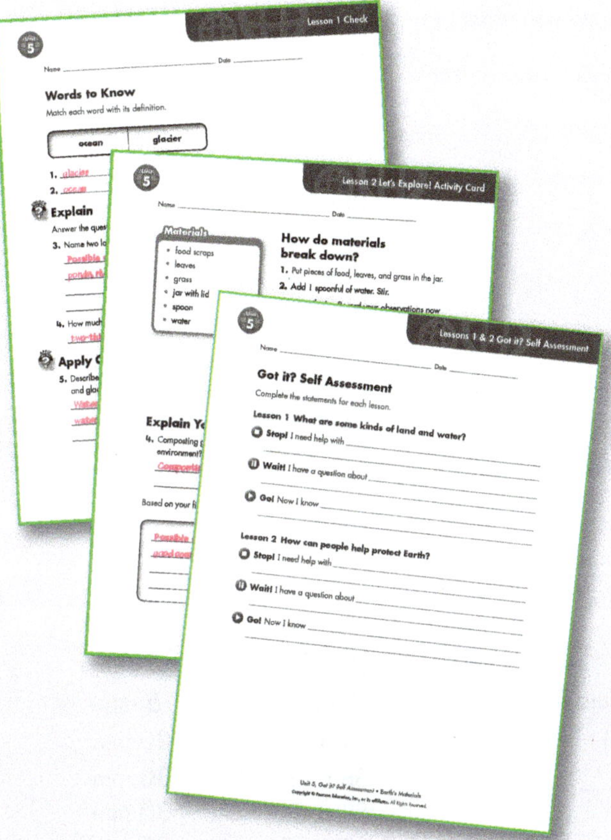

EVALUATE

There are numerous instances for evaluating learning and progress. At the end of each lesson, students can do a *Lesson Check*, and, at the end of each unit, they can watch the *Got it? 60-Second Video* to review key concepts. There is also a unit review, with targeted review strategies to address challenges students may have with the unit's content. *Got it? Self Assessments* help students to assess their progress and to judge what they need to study further. There are also *Got it? Quizzes* to help evaluate understanding of each unit. Unit-specific Study Guides and Concept Maps provide clear summaries and additional tools for evaluation and review.

21ST CENTURY SKILLS

The 21st Century skills of critical thinking, collaboration, communication, and creativity are methodically developed across the digital and print components for each level of *Shake Up Science*. In an increasingly globally competitive workforce, it is more critical than ever to prepare students for the careers of tomorrow.

COMMUNICATION OPPORTUNITIES

Activities in *Shake Up Science* are highly participative and require students to collaborate and share their ideas. There are a number of opportunities in each unit for students to communicate in pairwork, groupwork, or whole class activities, to give presentations, and to express themselves through writing.

ACCOUNTABILITY AND SELF-DIRECTION

Lesson Checks at the end of each lesson and *Got it? Self Assessments* at the end of each unit encourage students to self-evaluate and to make their own judgments about what they need to review.

CRITICAL THINKING AND PROBLEM SOLVING

Shake Up Science systematically cultivates students' skills of critical thinking and problem solving, with a science-related Big Question to lead each unit's learning, lots of exposure to scientific methodology, and *Think!* boxes relating to real-life topics.

DIGITAL LITERACY

Digital activities work hand-in-hand along with the print materials to help engage students and expand their understanding of scientific concepts as well as for review and feedback. Digital activities can be used in various ways during class to explore key scientific concepts.

Unit 1 — The Nature of Science

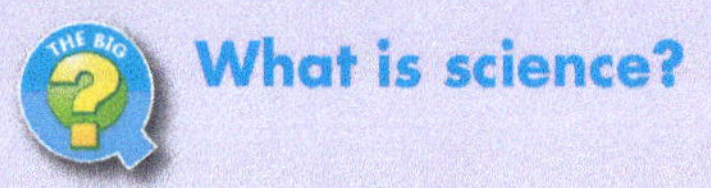

Lesson Plan

Unit Opener & Lesson 1 What questions do scientists ask?			
	Activity	**Pages**	**Time**
Engage	• Unit Opener: Think! *What is the girl doing?*	SB p. 4	5 min
	• Unit Opener: Things that help us observe.	SB p. 4	10 min
	• Unit Opener: Comparing things.	SB p. 4	10 min
	• Think! *Pretend you are a scientist. What animal do you want to study? Why?*	SB p. 7	10 min
Explain	• How scientists work together and observe objects	SB p. 5	20 min
	• Questions scientists ask	SB p. 6	20 min
	• More questions scientists ask	SB p. 7	20 min
Elaborate	• Describe Seeds	TB p. 5	20 min
	• Questions and Answers	TB p. 6	30 min
	• Kangaroos and Frogs	TB p. 6	15 min
Evaluate	• *Lesson 1 Check* (ActiveTeach)	TB p. 15a	10 min
	• Assessment for Learning	TB p. 7	10 min
	• Review (Lesson 1)	SB p 15	10 min
	• *Got it? Self Assessment* (ActiveTeach)	TB p. 15b	10 min
	• *Got it? Quiz* (ActiveTeach)	TB p. 15b	10 min

Lesson 2 How do scientists observe?			
	Activity	**Pages**	**Time**
Engage	• Think! *Is a pencil a tool?*	TB p. 9	5 min
	• Think! *Are all fish alike?*	TB p. 10	5 min
	• Think! *What are some school rules and home rules?*	TB p. 11	5 min
Explore	• Digital Lab: *How do we observe?* (ActiveTeach)	TB p. 8	30 min
Explain	• The five senses and observation	SB p. 8	30 min
	• Tools and measuring	SB p. 9	30 min
	• Comparing	SB p.10	30 min
	• Grouping and safety	SB p. 11	30 min
Elaborate	• Observe and Describe	TB p. 8	30 min
	• Apples and Orange	TB p. 9	20 min
	• Let's compare our fish!	TB p. 10	20 min
	• At-Home Lab: Group Objects	SB p. 11	20 min
	• Card Sort	TB p. 11	20 min
Evaluate	• *Lesson 2 Check* (ActiveTeach)	TB p. 15a	10 min
	• Assessment for Learning	TB p. 11	10 min
	• Review (Lesson 2)	SB p. 15	10 min
	• *Got it? Self Assessment* (ActiveTeach)	TB p. 15b	10 min
	• *Got it? Quiz* (ActiveTeach)	TB p. 15b	10 min

	Activity	Pages	Time
Engage	• Think! *What do scientists do?*	TB p. 13	10 min
Explain	• Testing ideas	SB p. 12	30 min
	• Collecting, recording, and sharing data	SB p. 13	30 min
	• *Got it? 60-Second Video* (ActiveTeach)	TB p. 13	10 min
Elaborate	• Flash Lab: Do a Test and Record Data	TB p. 12	20 min
	• Observe!	TB p. 12	20 min
Evaluate	• *Lesson 3 Check* (ActiveTeach)	TB p. 15a	10 min
	• Assessment for Learning	TB p. 13	10 min
	• Review (Lesson 3)	SB p. 15	10 min
	• *Got it? Self Assessment* (ActiveTeach)	TB p. 15b	10 min
	• *Got it? Quiz* (ActiveTeach)	TB p. 15b	10 min
Lab	• *Let's Investigate! How do things look?* (ActiveTeach)	SB p. 14	30 min

Flash Cards

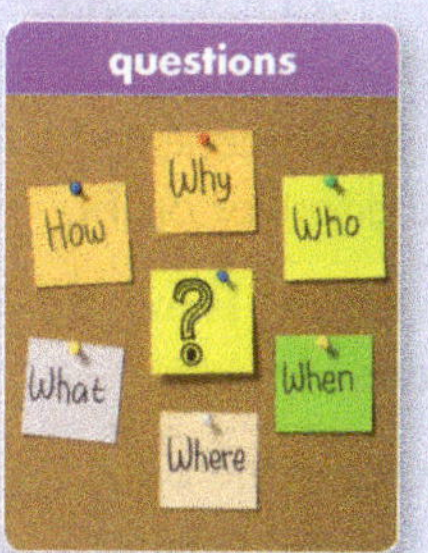

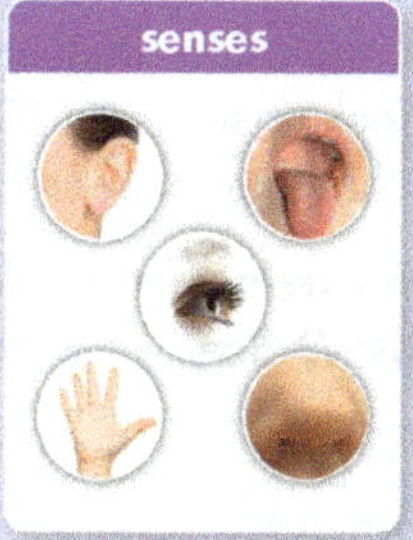

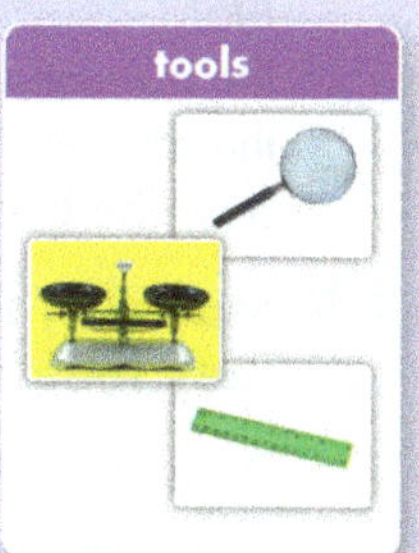

Lesson 1

Key Words	ELL Support
scientist, science, observe, objects, questions, answers	**Vocabulary:** *learn, hand lens, size, shape, color, seeds, diver, ocean, frog, baby, kangaroo, pouch, move, jump, firefly, fireflies, glow, light up, monkey, fish, bird, cat, dog, pretend* **Wh- Questions:** *what, where, why*

Lesson 2

Key Words	ELL Support
senses, tools, measure, compare, group	**Vocabulary:** *sound, bird, dog, fish, frog, sandwich, feather, shell, block, rock, bean, pattern, hand lens, ruler, balance, better, weigh, more, butterfly, alike, different, big, small, safety, safe, rules, goggles* **Verb:** *be*

Lesson 3

Key Words	ELL Support
collect, data, record (v), chart	**Vocabulary:** *words, pictures, numbers, mark, soak up, follow*

Unit 1 — The Nature of Science

Unit Objectives

Lesson 1: Students will learn that scientists ask questions to learn.

Lesson 2: Students will learn ways scientists observe things.

Lesson 3: Students will learn how scientists collect and record data.

Vocabulary: *science, scientist, observe, ruler, hand lens, clock, observe, size, bird, alike, flower, grass, beak, tail*

Introduce the Big Question

What is science?

Build Background Say *What time is it?* Look at your watch and write the time on the board. Have students gather around you. Hold up a piece of paper or book. Say *How long is it? Let's observe!* Take out a ruler and measure. Record the measurement. Then say *What does it look like?* Look at a piece of paper or a page of a book through a hand lens. Say and write *I can see (the letters).* Finally, say *What time is it?* Again, look at your watch and record the new time. *What did we find out? How big (the paper) is! What it looks like!* Point to the times. *How long did it take us to observe these things? It took us (five) minutes!*

Engage

Think!

What is the girl doing?

Draw students' attention to the picture. Read the question aloud. Allow students time to discuss what they think the girl is doing. Provide support as needed.

1 **Circle what you can use to see things.**

Point to the pictures and elicit or say the names of the items. *What can you do with a ruler? What can you do with a hand lens? What can you do with a clock?* (Possible answers: *See how (long) things are. Look at things. Tell the time.*) Read the question aloud and invite students to circle the answer. Check answers as a class.

2 **Circle the part of your body you can use to observe the color of a bird.**

Invite students to identify the body parts pictured and the senses associated with each body part. *What*

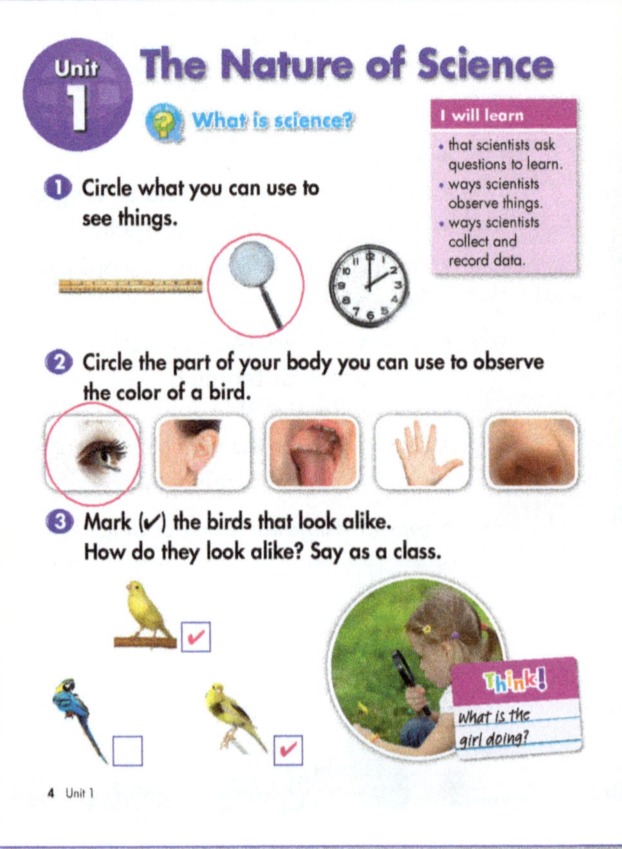

can you do with your eyes? See! Then ask questions to solicit what sense can tell students about a bird's color. *Do your hands tell you what color a bird is? No! Can you smell what color a bird is? No!* Read the instructions and allow students time to circle the answer. Check answers and invite students to describe birds that they've seen outdoors or on TV.

3 **Mark (✔) the birds that look alike. How do they look alike? Say as a class.**

Draw students' attention to the pictures of the birds. *Have you ever seen birds like these ones? Where? What did they look like?* Provide vocabulary support as needed. Point out similarities and differences. *What color are these two birds? Yellow! This bird has some black on it. This bird does, too.*

Draw students' attention to the shapes and sizes of the birds as well. *Look at this bird. It has a long tail! Its beak has a different shape, too.*

Ask students to mark which birds are alike. Invite volunteers to say which birds they picked and encourage them to explain how the birds they picked look alike. Accept all logical answers

Think! Again!

Revisit the question *What is the girl doing?* Invite students to share their ideas freely. (Possible answers: *She is looking at the flower. She is observing the flower petals.*) Provide vocabulary support as needed and accept all logical answers.

What questions do scientists ask?

Objective: Learn that scientists ask questions and observe.

Vocabulary: *scientist, science, observe, objects, size, shape, color, seeds, diver, ocean, tool*

Digital Resources: Flash Cards (*scientist, observe, objects*)

Materials: hand lens (1 per small group), sets of small edible seeds of different shapes, sizes, and colors (chia, sesame, amaranth, poppy, cumin, rice, etc.)

Unlock the Big Question

Write the following text on the board: *I will learn how scientists work. What questions do they ask? How do they observe?*

ELL Vocabulary Support

Review or pre-teach *hand lens* and *seeds*. Hold up a hand lens. *What is this? It's a tool that helps us see things! What are seeds? Plants grow from seeds.* Explain that not all seeds are edible.

Build Background Set out different kinds of seeds in separate areas. Divide the class into the same number of groups as kinds of seeds. Distribute a hand lens to each group. Each group starts with a different seed and students take turns examining the seeds. *Does the hand lens help you see the seeds? What do the seeds look like?* Explain to students that they just did something scientists do and that they will learn more about what scientists do.

Explain

1 Read. What does a scientist do? Say as a class.

Show the *scientist* Flash Card and elicit the word. Read the paragraph aloud for students and ask the question. Guide students to answer. *Scientists learn about the world around us.*

2 Do scientists work together? Say with a partner.

Read the paragraph aloud again. Pair students and have them discuss the answer. Ask them to point to where in the paragraph or on the page they learn that scientists can work together. Draw students' attention to the photo of the two scientists. Ask students to describe what they are doing. Provide vocabulary support as necessary and accept all logical answers.

Lesson 1 · What questions do scientists ask?

Key Words
- scientist
- science
- observe
- objects
- questions
- answers

1 Read. What does a scientist do? Say as a class.

Science and Scientists

A **scientist** uses **science** to learn about the world around us. A scientist can work with other scientists. They learn new things together. You can use science to learn, too.

2 Do scientists work together? Say with a partner.

3 Read. Mark (✔) the scientists who observe things.

Observe

Scientists observe. **Observe** means to find out about things. You can observe the size, shape, and color of **objects**. You can observe other things, too.

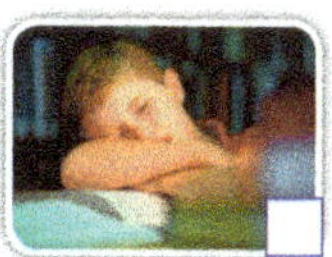
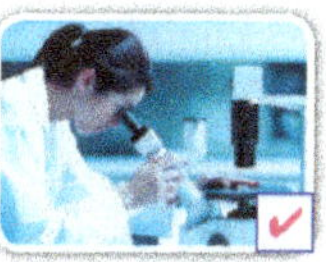

3 Read. Mark (✔) the scientists who observe things.

Display the *observe* and *objects* Flash Cards. Ask students to describe what they see. Then draw their attention to the three photos. Elicit from students or describe what the person in each photo is doing. *Look! This person is underwater. It's a diver! The diver observes fish, coral, and other animals.*

Read the paragraph. Elicit from students what *observe* can mean. (*To find out about things.*) Read the instructions and have students mark the pictures. Check answers as class and discuss students' answers.

Elaborate

Describe Seeds

Have students get in their groups from the beginning of class and examine the seeds again, following the same procedure. Write the names of each kind of seed on the board. *Let's observe the seeds!* Ask questions *How big are the seeds? What color are they? Are they the same shape as the other seeds?*

Encourage students to describe or draw and compare with one another the color, size, and shape of each kind of seed. Record on the board some of the descriptions of each kind of seed. It does not matter whether students describe the seeds correctly or identify them, only that students engage in the process of observing and comparing. *We are asking questions and observing. We found out that (sesame) seeds have a different shape and color than (rice). You're little scientists!*

What questions do scientists ask?

Objective: Learn what kind of questions scientists ask and answer.

Vocabulary: *questions, answers, animals, frog, kangaroo, pouch, what, where, why, how, move, jump*

Digital Resources: Flash Card (*questions*), *I Will Know...* Digital Activity

Build Background Write *Question* and *Answer* on the board. Show the *questions* Flash Card and explain. *A question is something we ask. For example,* How old are you? *We ask questions about people, places, and objects.* Elicit from students other things we ask questions about. (Possible answers: *the weather, food, people's feelings,* etc.) *An answer is what we find out. When we get answers to our questions, we learn things!* Elicit from students a question and write it under *Question*. Point out the question mark. Then elicit an answer and write it under *Answer*. (*What color is the sky? It is blue.*)

Explain

4 Look at the leaves. What can you say about them? Say with a partner.

Pair students. Ask them to look carefully at the photos and to describe the leaves in each photo. Invite volunteers to share their answers with the class. (Possible answers: *They're (red). They're (small).*)

ELL Vocabulary Support

Point out the pictures on the right-hand side of the page. Elicit or identify for students the plants and animals: *pine tree, frog, kangaroo.* Point out that the kangaroo has a baby. *It keeps its baby in its pouch!*

5 Read. Match the questions and answers with the pictures.

Read the paragraph aloud. Explain the instructions and then read the questions and answers. Elicit the question and answer in each instance and write them under the headings on the board.

Remind students that scientists ask questions and observe to get answers. Invite students to match the question and answer pairs to the pictures.

Ask questions to check answers. *What does question a ask about? It asks about a plant. Which picture shows a plant? The first picture. What is the answer? It's a tree!*

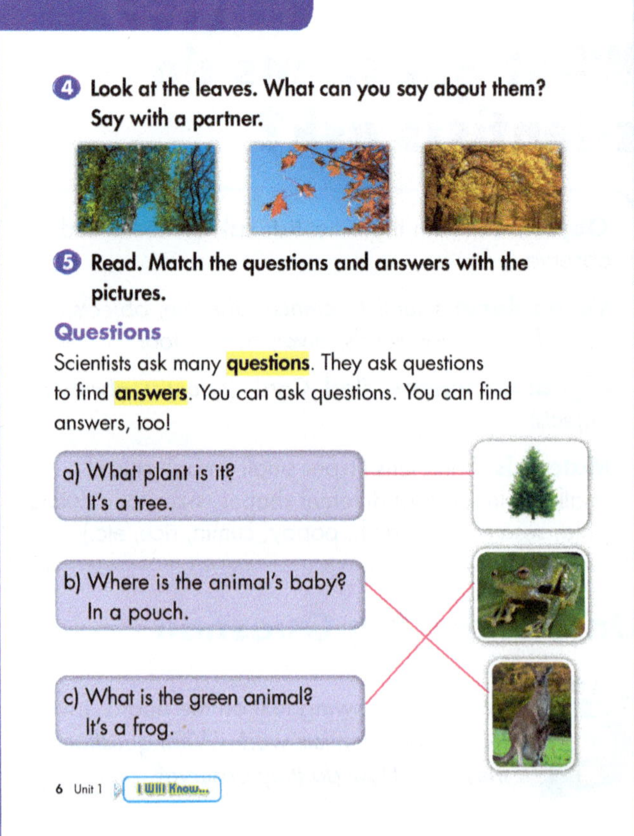

ELL Language Support

Point to the questions on the board. Underline *What color, Where is,* and *What kind.* Explain that we can use these words to ask questions. You may wish to add *Why* and *How* questions as well. You may also wish to point out the rising inflection at the end of questions.

Elaborate

Questions and Answers

Divide the class into small groups. Refer students back to the pictures for exercise 5. *What other questions can we ask?* Encourage groups to brainstorm. Remind students they can use the samples on the board to help start new questions. Monitor and provide support as needed. (Possible questions: *What color is the tree? What is the animal in the picture? How big is the frog? Why does the frog have big eyes?*)

Kangaroos and Frogs

Write *How do kangaroos and frogs move?* on the board. Take students outside. Divide the class into Kangaroos and Frogs. Kangaroos hold their arms loosely in front, like the kangaroo in the picture, and jump using both feet when you call out *Kangaroos.* Frogs get down on all fours and "jump" when you say *Frogs.* Groups switch animals and repeat. Finally, return to the classroom and, as the answer to the question, write *Frogs and kangaroos jump.*

I Will Know...

Have students do the *I Will Know...* Digital Activity.

What questions do scientists ask?

Objective: Learn more about questions scientists ask.

Vocabulary: *what, where, why, firefly, fireflies, glow, light up, light, monkey, frog, kangaroo, fish, bird, cat, dog, pretend*

Digital Resources: Flash Card (*questions*), *Lesson 1 Check* (print out 1 per student)

Materials: Animal Cards (selections, including *cat, dog, firefly, fish, frog, kangaroo, monkey,* and *parrot/bird*)

Build Background Put the Animal Cards face down on a table. Invite volunteers to come up, one at a time, turn over a card, and mime the animal for the class. The class guesses the animal. Invite volunteers to say something about each animal. *It's a bird/parrot. A bird/parrot has feathers. It can fly. It's a cat. A cat says meow. It's a firefly. Fireflies glow, or light up.* Provide support as necessary.

Explain

6 **Look at the pictures. What are three questions the boy can ask about the animals? Say as a class.**

Point to the first picture and have students read the questions. Then draw students' attention to the second and third pictures. *What is in the third picture? A firefly. What does the boy have in the jar? Fireflies!*

Next, read the question and solicit ideas. Write students' ideas on the board. Accept all logical questions, but encourage students to include a *What*, a *Where*, and a *Why* question. If students do not come up with it on their own, ask *Why do fireflies glow?* Explain to students that fireflies can make their own light. *Other fireflies can see them!*

ELL Content Support

Fireflies are a kind of beetle, and there are more than 2,000 species of fireflies. Some of these species flash lights at night, which can be yellow, green, or orange. They do this to attract each other or to protect themselves from predators. Other creatures, such as deep-water fish, can also create their own light.

7 **Look at the monkey. Say two questions you can ask with a partner.**

Draw students' attention to the picture of the monkey and have them identify the animal. Pair students and read the instructions. Have students brainstorm

questions they can ask. Monitor and provide support. If necessary, point to the questions the girl asks in the photo at the top of the page to help students form questions. (*Where is the monkey? What does the monkey do? What color is the monkey?*)

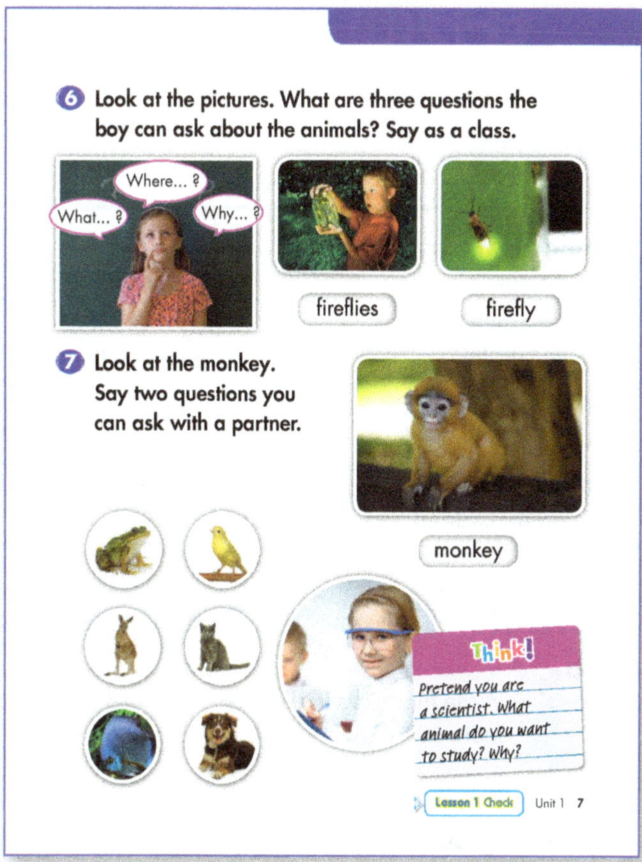

6 Look at the pictures. What are three questions the boy can ask about the animals? Say as a class.

7 Look at the monkey. Say two questions you can ask with a partner.

Think!

Pretend you are a scientist. What animal do you want to study? Why?

Draw students' attention to the photo of the girl and read the question. Point to the pictures on the left and elicit or identify the animals. Ask students whether they would like to study these or any other animals. *What do you want to know about the animal?* Allow time for all students to answer, and encourage each student to ask a question about each animal that they'd like to study.
(*I want to study frogs. Where do they live? I want to study kangaroos. How high can they jump?*) Provide support as necessary and accept all logical answers.

Evaluate

Lesson 1 Check   Assessment for Learning

Review the Key Words for Lesson 1 (see Student's Book page 5). Distribute the *Lesson 1 Check* and guide students as they complete it. Check answers as a class. Then ask students to grade their progress on the topic of what questions scientists ask from 1 to 3: 3 = *I understand what questions scientists ask;* 2 = *I need to study more;* 1 = *I need help!* Encourage students giving themselves a 1 or a 2 to say what they found difficult and what they need to study more.

Lesson 2

How do scientists observe?

> **Objective:** Learn about observing using the five senses.
>
> **Vocabulary:** *senses, sound, bird, dog, fish, frog, sandwich, feather, shell, block, rock, bean, pattern*
>
> **Digital Resources:** Flash Card (*senses*), *Let's Explore!* Digital Lab
>
> **Materials:** shell, wooden block, feather, patterned paper, rock, dried bean, hand lens, bell, fruit

Unlock the Big Question

Write the following text on the board: *I will learn some ways scientists observe.*

Build Background Display the *senses* Flash Card. Invite students to identify each body part and the sense associated with it. Provide support as needed.

Explore

Let's Explore! Lab How do we observe?

Objective: Learn some ways to observe and describe objects.

Digital Resources: *Let's Explore!* Digital Lab, *Let's Explore! Activity Card* (1 per student)

- Gather students around a desk or table. Display the items and identify them. Model how to observe objects by looking, feeling, smelling, and hearing. Drop some on the table to see if they make a noise.

- Invite a volunteer to pick up one of the items, and ask questions that elicit descriptions. *What color is it? What size is it? Is it big or small? How does it feel? Is it (soft)? What sounds does it make?* Provide vocabulary support as necessary and write some descriptors on the board next to the name of the corresponding item. Repeat with the other objects.

- Show the Digital Lab and invite groups to do the activity. Remind students they can refer to information on the board as they go through the activity.

- Have students work in pairs to complete the *Activity Card*.

Explain

1 **Read. Look at the fish. What colors do you see?**

Read the paragraph aloud. Draw students' attention to the fish and ask the question. Accept all logical answers.

Lesson 2 · How do scientists observe?

1 **Read. Look at the fish. What colors do you see?**

Senses

Scientists use their **senses** to observe. You can use your senses, too. You look to observe things like size, shape, and color. You listen to observe sounds.

2 Point to the big fish. Point to the small fish. What fish do you like more? Why?

3 Look around the classroom. Say three objects you see.

4 Circle the things you can hear.

Key Words
- senses
- tools
- measure
- compare
- group

8 Unit 1 Let's Explore! Lab

---

2 **Point to the big fish. Point to the small fish. What fish do you like more? Why?**

What sizes are the fish? One is big, and one is small! Invite students to point accordingly and answer which they like more. Encourage students to explain why they answered the way they did. (*I like orange. It's big.*) Provide support as necessary.

3 **Look around the class. Say three objects you see.**

Invite students to look around and name three objects. Invite volunteers to share. *(John), what can you see? I see a pencil.* Write the names of some of the objects on the board and practice the words.

4 **Circle the things you can hear.**

What do you hear with? My ears! Invite volunteers to identify the items pictured by reading the labels. *Can you hear a dog? Yes! Can you hear a sandwich? No!*

Elaborate

Observe and Describe

Have students review exercise 2, page 4. Gather students around again. Invite a volunteer to ring the bell. *This is a bell. It makes a sound. What body part can you use to hear sounds?* Have students say or point to the appropriate picture(s) in the book. Elicit or provide descriptions. *What is the sound like? The bell makes a nice sound. Do you like it?* Provide support as necessary. Repeat with the other items to emphasize which senses we can use to describe each one. End with the fruit. Guide students to notice that they can see, smell, touch, and taste the fruit.

How do scientists observe?

Objective: Learn about some tools for observing things.

Vocabulary: *tools, hand lens, ruler, balance, measure* (v), *sandwich, better, weigh, more*

Digital Resources: Flash Cards (*tools, measure*), *I Will Know…* Digital Activity

Materials: 2 apples, 1 orange, balance

Build Background Display the *tools* and *measure* Flash Cards. *What tools can you see? What do they help us do?* Elicit plausible answers. Explain to students that scientists can use tools to observe things, too. *Scientists use their senses to observe things. They can use a hand lens to help them see things better.* Explain that a tool is anything that helps you do something and that some tools can help you observe.

Explain

5   Read. Circle the tools.

Read the paragraph aloud for students. Write the names of the tools on the board and say them aloud for students to repeat. Then have students circle the tools. Check answers as a class.

Ask questions to check comprehension. *What is a tool? Something that helps us do something, like observe. Say one tool.* (*Balance.*) *What does measure mean? To find out how much, how long, or how tall.*

6   Say as a class. Match the tools to the questions

Invite volunteers to read each word in the colored boxes. Read the first speech bubble aloud and ask students to shout out which tool matches the question. (Answer: *Balance.*) Repeat for the remaining words and speech bubbles. Finally, have students match the words to the questions in their books.

Write *Tools* on the board. Give a definition of *tool* and check understanding. Then practice some of the new vocabulary. *Some tools help us to observe. To observe is to find out about things. Can you name a tool that helps us to see? A hand lens! Right!* Write *hand lens* under *Tools. What is a tool that helps us to measure? A* (*ruler*). Add (*ruler*).
Challenge students to name as many tools as they can think of and to say or mime what they can help us do. *Right! A crayon helps us to draw. A pencil helps us to write.*

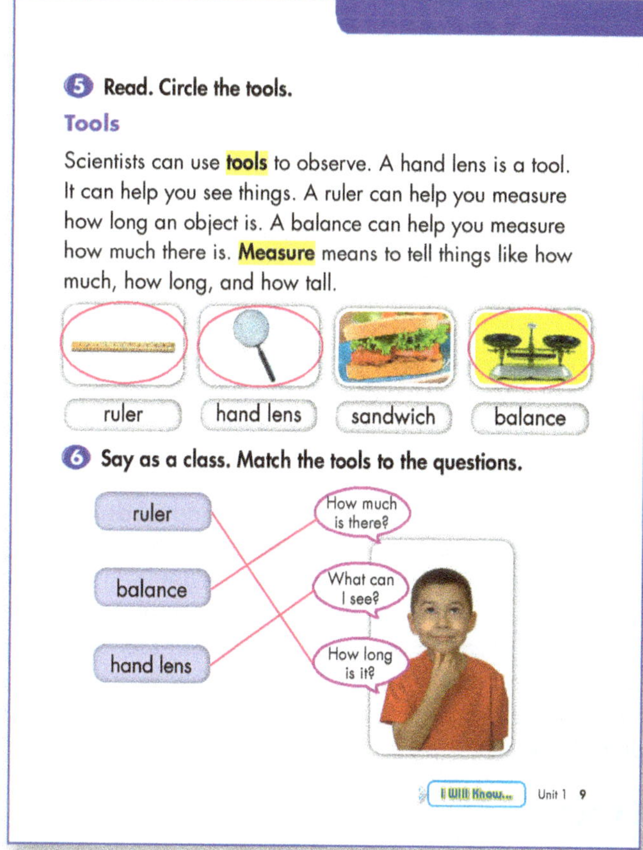

Elaborate

Apples and Orange

Gather students around a desk or table. Display the balance and have students identify it and what it does. (*Balance. It tells how much, or weighs things.*) Display the fruit and elicit their names or identify them for students. *We're going to weigh the apples and orange. Which do you think weighs more?* Allow students to hold the fruit and make guesses. Write *apples* and *orange* on the board. Weigh the fruit and record the weights under the appropriate word. *How much do the apples weigh? How much does the orange weigh?* Record the relevant weights. *Were you right?*

Think!

Is a pencil a tool?

Remind students how they use a pencil and what it helps them do. Then ask the question and guide students to answer that a pencil is a tool we can use to write things.

I Will Know…
Have students do the *I Will Know…* Digital Activity.

Lesson 2

How do scientists observe?

Objective: Learn how to compare.

Vocabulary: *butterfly, compare, alike, different, big, small*

Digital Resources: Flash Cards (*tools, measure, compare*), Animal Cards (*fish* and *goldfish*; make copies so each student has one card)

Build Background Display the *tools* and *measure* Flash Cards and review with students what they have learned about how scientists observe. *What is one way scientists observe things? They use their senses. What is another way they observe things? They can use tools to observe, too. We're going to learn another way scientists can observe things. They can say how things are alike and different!*

Explain

7 **Circle *T* (true) or *F* (false).**

Read each sentence for students and allow time for them to circle the answers. Check answers and correct the false statement as a class. (*You have five senses.*)

8 **Read. Look at the picture. How are the fish alike? Say with a partner.**

Display the *compare* Flash Card and ask students to discuss the similarities and differences between the apple and the orange. Have students read the paragraph silently. Write *compare* on the board and ask volunteers to say what it means. (*To look at the similarities and differences between two items.*) Then draw students' attention to the photo of the fish. *How many fish are there? Two. How are they alike?* Have pairs answer the question and share their answers with another pair. (Possible answers: *They are yellow. They are the same color. They are about the same size.*)

9 **Look at the butterflies. Compare. Say as a class.**

Point to the butterflies. *How are they alike? How are they different?* (Possible answers: *They are butterflies. They have the same shape. They are different colors.*) Ensure students notice both how the butterflies are alike and how they are different.

7 Circle *T* (true) or *F* (false).

1. Scientists use tools to observe. (T)/ F
2. You can observe how big or small something is. (T)/ F
3. You only have three senses. T /(F)

8 Read. Look at the picture. How are the fish alike? Say with a partner. 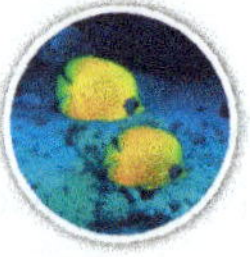

Compare

Scientists say how things are alike. They say how things are different. **Compare** means to say how things are alike and different.

9 Look at the butterflies. Compare. Say as a class.

Help students review the verb *be*. Write *is* and *are* on the board. Say sentences omitting the verb *be* for students to call out *is* or *are*. For example, *This butterfly (beep) blue.* Students should call out *Is! The two butterflies (beep) pretty.* Students should call out *Are!*

Elaborate

Let's compare our fish!

Distribute an Animal Card to each student. Invite students to color their fish. Encourage them to make patterns, e.g., stripes, spots, etc. Finally, put students into small groups to compare their fish. *My fish is big. My fish is blue and yellow. My fish has red spots.* Monitor and provide support as needed.

Think! 

Are all fish alike?

Read the question aloud and invite students to answer freely. Guide students to conclude that fish are alike–they are fish–but also different–they can be different sizes, colors, and so on.

Lesson 2
How do scientists observe?

> **Objective:** Learn how scientists group things.
>
> **Vocabulary:** group (v), group (n), *fish, butterfly, safety, safe, rules, goggles*
>
> **Digital Resources:** *Lesson 2 Check* (print out 1 per student)
>
> **Materials:** vocabulary cards (1 set per group; write the following words on index cards: *balance, ruler, hand lens, scientist, senses, safety, fish, butterfly, frog*), safety goggles (1 per small group)

Build Background Divide the class into groups according to a similarity. For example, you could ask all the boys to stand on one side of the room and all the girls to stand on the other. Have students notice the points of similarity and difference between the groups.

Explain

10 **Read. Circle the things that are alike.**

Read the paragraph aloud. *We just made groups of (all girls and all boys)!* Turn students' attention to the photos. *What is in the first photo? A fish. What is in the second photo? A fish. What is in the third photo? A butterfly.* Invite students to circle the items that are alike. *Right! They are fish!* Ensure students understand that, by comparing things, we can say if they are alike and put them into groups accordingly.

11 **Read. Circle the things that help you stay safe.**

Read the paragraph aloud. Invite volunteers to read the labels. Then read the list of rules and guide students to understand them. Remind students that a tool is anything that helps you do something. Divide the class into small groups and distribute safety goggles. Have students take turns putting them on. *What does a ruler do? It helps us tell how long or tall something is. Does it help you stay safe? No! What do safety goggles do? Keep my eyes safe! What else helps you stay safe in science class? Rules!*

Think!

What are some school rules and home rules?
Brainstorm school rules that help students stay safe. Allow students to mime as needed. Accept all logical answers and write them or draw pictures on the board. Do the same with home rules. Then hold a class vote for the most important safety rules. (*Don't run down the stairs. Don't open the windows. Don't push each other.*)

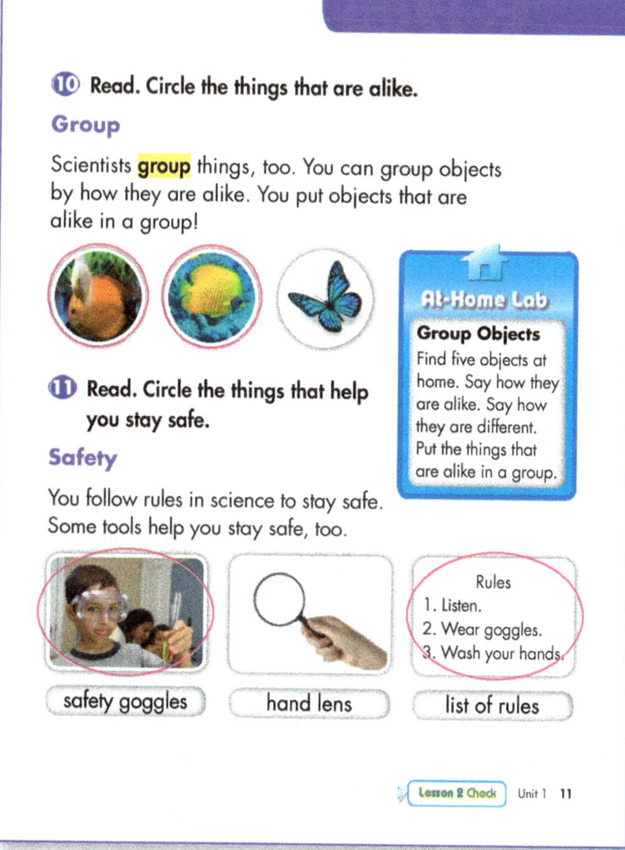

Elaborate

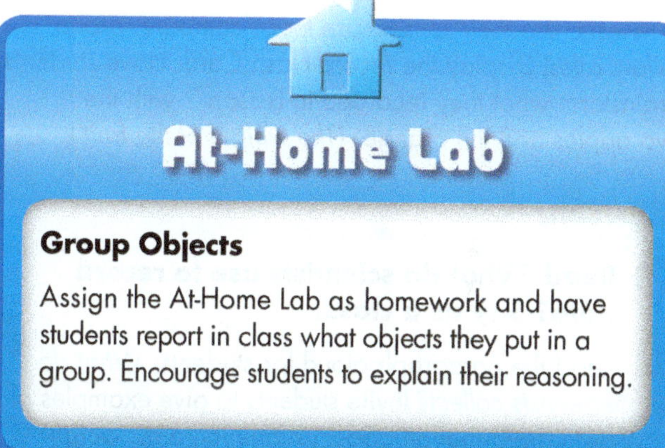

Card Sort

Divide the class into groups and distribute sets of index cards, mixed up. Challenge groups to sort the cards into groups (e.g., animals, tools, words that start with *s*).

Evaluate

Lesson 2 Check   **Assessment for Learning**

Review the Key Words for Lesson 2 (see Student's Book page 8). Distribute the *Lesson 2 Check* and guide students as they complete it. Check answers as a class. Then ask students to grade their progress on the topic of ways scientists observe from 1 to 3: 3 = *I understand ways scientists observe*; 2 = *I need to study more*; 1 = *I need help!* Encourage students giving themselves a 1 or a 2 to say what they found difficult and what they need to study more.

How do scientists collect and record data?

Objective: Learn how scientists collect and record data.

Vocabulary: *collect, data, record, chart*

Digital Resources: Flash Card (*record*), *I Will Know...* Digital Activity, Animal Cards (1 copy of each of the following: *cricket, firefly, grasshopper, ladybug, mosquito, moth, slug, snail, spider, worm, ant, bee, beetle*)

Materials: per small group: cup with a small amount of water in it, paper towels, tin foil, sponge, pieces of paper

Unlock the Big Question

Write the following on the board: *I will learn how scientists collect and record data.*

Build Background Review the ways scientists observe that students have learned about so far in this lesson. (Answers: *use their senses, use tools, compare, group, follow rules*) Display the *record* Flash Card. Have students brainstorm what they recorded in Lesson 2 with the two apples and the orange. (*The weight of the fruit.*)

Explain

1 Read. What do scientists use to record data? Say as a class.

Read the paragraph aloud for students. *What do scientists collect?* Invite students to give examples of data. (*Information about an animal, e.g., weight, color, abilities, etc.*) Elicit how scientists collect data and draw an example of a chart on the board.

2 Look at the picture. Draw the animal the girl is observing.

Draw students' attention to the picture and elicit the animal. (*snail*) *What do you think the girl is observing? What can she record about the snail?* Accept all logical answers. (Possible answers: *She's observing the color and shape of the snail.*)

Give students time to draw the snail. Invite volunteers to show their drawings to the class.

I Will Know...

Have students do the *I Will Know...* Digital Activity.

Lesson 3 · How do scientists collect and record data?

1 Read. What do scientists use to record data? Say as a class.

Scientists Collect and Record Data

Scientists **collect** information. In science, information is called **data**. Scientists **record** data. They can use words, pictures, numbers, or **charts**.

2 Look at the picture. Draw the animal the girl is observing.

Key Words
• collect
• data
• record
• chart

12 Unit 1   I Will Know...

Elaborate

⚡ Flash Lab

Do a Test and Record Data

Divide the class into small groups and distribute materials. *We're going to do a test and record what we find!* Draw a simple chart on the board with the materials and ask students to copy it. *Do paper towels soak up water?* Invite students to say whether they think the answer is *yes* or *no*. Then invite groups to do the test. *Were you right? Yes!* Invite students to mark (✔) or (✗) accordingly on their charts. Ask students to test the other materials and record the information. Check answers as a class.

Observe!

Display the Animal Cards around the class. Ask students to walk around the class in pairs and observe the animals one at a time. Encourage students to discuss what they find interesting about each animal. Monitor and provide support as necessary.

Ask students to sit down in pairs and decide which two animals they want to learn about more. When they're ready, ask students to look at the two animals again. Have students ask questions and discuss similarities and differences. Provide vocabulary as necessary.

How do scientists collect and record data?

> **Objective:** Learn how scientists collect and record data.
>
> **Vocabulary:** *collect, record, data*
>
> **Digital Resources:** Flash Card (*record*), *Lesson 3 Check* (print out 1 per student), *Got it? 60-Second Video*

Build Background Write students' names on the board. Check attendance, and write (✔) beside each student who is present and (✘) beside those who are not. Explain that you collect information about who is at school and record that information. Tell students they will learn about how scientists collect and record data in today's lesson.

Explain

3 **Read. What can a mark in a chart show? Say with a partner.**

Read the paragraph aloud for students. Have students say how scientists collect data and what a mark in a chart can mean. Ask them to think about the Flash Lab test and what the marks meant in their charts. (*Whether the material soaks up water.*)

4 **Ask five friends, "Do you like dogs, cats, or birds best?" Mark (✔) each answer in the chart.**

Focus students' attention on the pictures of the animals and have them say the names. Then read the question aloud and invite students to answer. As they do so, demonstrate putting marks in the chart.

Then give students time to work in groups asking each other the question and adding marks.

5 **Count the marks for each animal. Which is your friends' favorite animal? Compare with other groups.**

When they finish, ask students in their groups to look at the marks and say their group's favorite animal. Ask them to share the information in class and explain why they like each animal.

Think!

What do scientists do?

Elicit from students some things that people do in science. (Possible answers: *ask questions, find answers, test ideas, use their senses to observe,* etc.) Read each item and ask volunteers to tell about times when they have done those things. Put a check mark next to the item. Help students understand that, in science, people work together to find

3 Read. What can a mark in a chart show? Say with a partner.

Collect and Record Data

You can collect data by asking questions. You can record data in a chart. For example, one mark in a chart can record one person's answer to a question.

4 Ask five friends, "Do you like dogs, cats, or birds best?" Mark (✔) each answer in the chart. Sample data

Favorite Animals					
🐱 cats		✔			
🐶 dogs	✔		✔		✔
🐦 birds				✔	

5 Count the marks for each animal. Which is your friends' favorite animal? Compare with other groups.

answers to their questions. They look for answers by doing tests, observing, measuring, and recording information.

Evaluate

Lesson 3 Check   **Assessment for Learning**

Review the Key Words for Lesson 3 (see Student's Book page 12). Distribute the *Lesson 3 Check* and guide students as they complete it. Check answers as a class. Then ask students to grade their progress on the topic of how scientists collect, record, and share data from 1 to 3: 3 = *I understand how scientists collect, record, and share data;* 2 = *I need to study more;* 1 = *I need help!* Encourage students giving themselves a 1 or a 2 to say what they found difficult and what they need to study more.

> **Got it?**   **60-Second Video**
>
> Play the *Got it? 60-Second Video* to review the unit material.

Let's Investigate!

In this unit, students learn some ways scientists observe. In this lab, they will investigate how adding water can change how we see objects through a viewer.

Let's Investigate! Lab **How do things look?**

Objective: Students will observe how water can change how things appear in a viewer.

Materials: one set per group: viewers, one with water, various objects, for example, a colored pencil, a shell, a plastic animal, a crayon

Digital Resources: *Let's Investigate!* Digital Lab, *Let's Investigate!* Activity Card (1 per student)

Advance Preparation: Prepare two viewers per group: cut two 10 cm holes in either side of empty paper paint buckets. Secure plastic wrap over the top of each with a rubber band. Pour some water on top of the plastic wrap of one of the viewers for each group. You may wish to wait until you've set up stations for groups to make the viewers with water.

- Explain that students will view some objects through the viewer without water. Then they will look at the same objects through the viewer with water and compare how the objects look.

- Divide the class into groups and distribute materials. Have students look at the first item through the viewer without water. Provide ample time for each student to look at the object in the viewer.

- Then invite students to look at the same object through the viewer with water. *How does the water change what you see?* Elicit that the water makes the object look bigger.

- *How is the viewer with water like a hand lens? It makes objects look bigger, too.*

- Have students complete the Activity Card.

Teacher Time-Saving Option: Show the *Let's Investigate!* Digital Lab as an alternative to the hands-on lab activity.

Unlock the Big Question

Have students refer to the Big Question on the Unit Opener page. In pairs have them recall what they have learned about what scientists do. Have pairs complete questions 4 and 5 on the *Activity Card*.

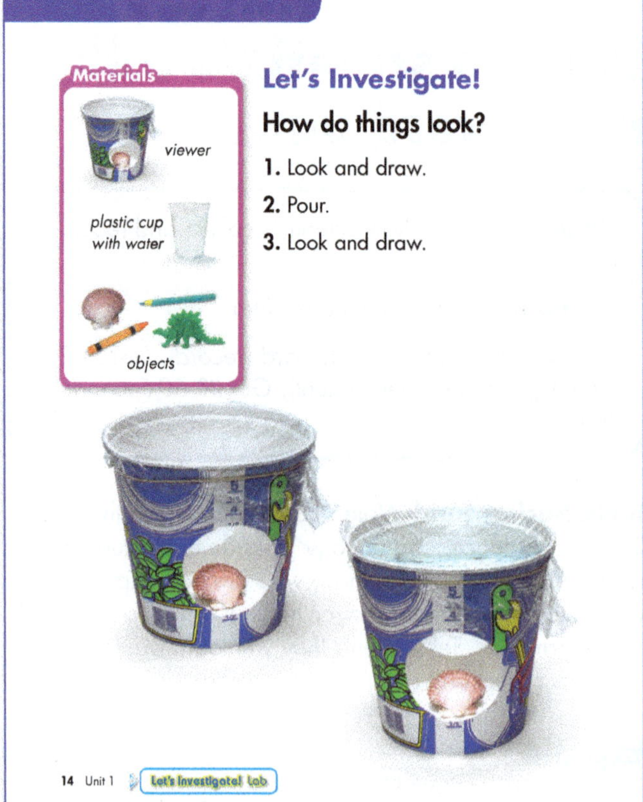

Let's Investigate!

How do things look?

1. Look and draw.

2. Pour.

3. Look and draw.

14 Unit 1 Let's Investigate! Lab

Class Project: Safe Science Collage

Materials: art supplies

Brainstorm with students a list of rules and tools that will help them stay safe in science class. Allow students to mime if they wish. Accept all logical answers and provide vocabulary support as necessary. Write a list on the board. (*Listen to the teacher. Be careful with tools. Wear safety goggles. Wash your hands.*) Then distribute art supplies and have students draw themselves following one of the rules or engaging in another safe practice in science class. Invite volunteers to share their drawings. Make a science safety collage and display it in the classroom.

Unit 1 Review

What is science?

Digital Resources: Print out 1 of each per student: *Got it? Self Assessment, Got it? Quiz*

Evaluate

Strategies for Targeted Review

The following are strategies for providing targeted review for students if they encounter challenges with the content.

Lesson 1 What questions do scientists ask?

Question 1

If... students are having difficulty answering the questions, then... direct students to review Lesson 1.

Lesson 2 How do scientists observe?

Question 2

If... students are having difficulty tracing the words, then... ask them to trace them slowly and help them with specific letters they find difficult to trace.

Lesson 3 How do scientists collect and record data?

Question 3

If... students are having difficulty matching, then... make a list as a class of things scientists do: *ask questions, find answers, test ideas, learn new things, use their senses to observe, use tools, record data, share what they learn, follow rules.* Then have students match the answers.

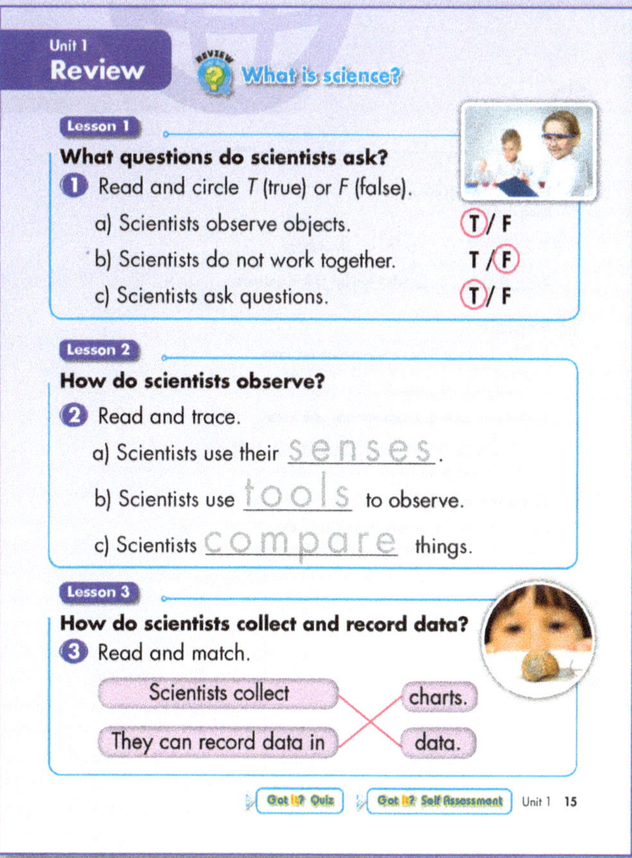

ELL Language Support

Before students start working on the Review activities, have them read each question aloud along with you.

Got it? Self Assessment

Immediately after students have completed the Review activities, distribute a *Got it? Self Assessment* to each student. Have students complete the *Stop! Wait!* and *Go!* statements for each lesson, allowing them to look back through the lesson material if necessary.

Got it? Quiz

Distribute a Unit 1 *Got it? Quiz* to each student. Quizzes may be used for assessing students' understanding of unit concepts as well as for grading purposes.

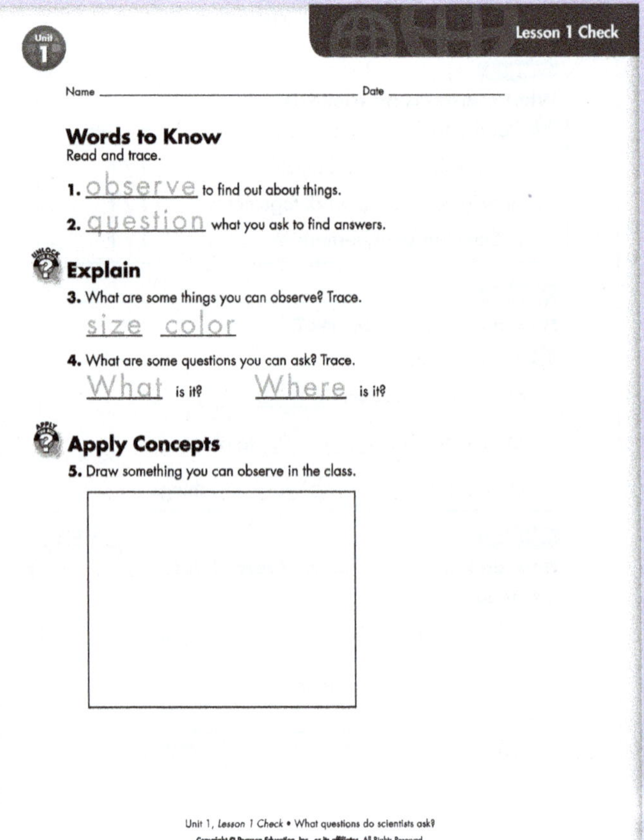

Name _____________________ Date _____________

Words to Know
Read and trace.

1. <u>observe</u> to find out about things.

2. <u>question</u> what you ask to find answers.

Explain

3. What are some things you can observe? Trace.

<u>size</u> <u>color</u>

4. What are some questions you can ask? Trace.

<u>What</u> is it?   <u>Where</u> is it?

Apply Concepts

5. Draw something you can observe in the class.

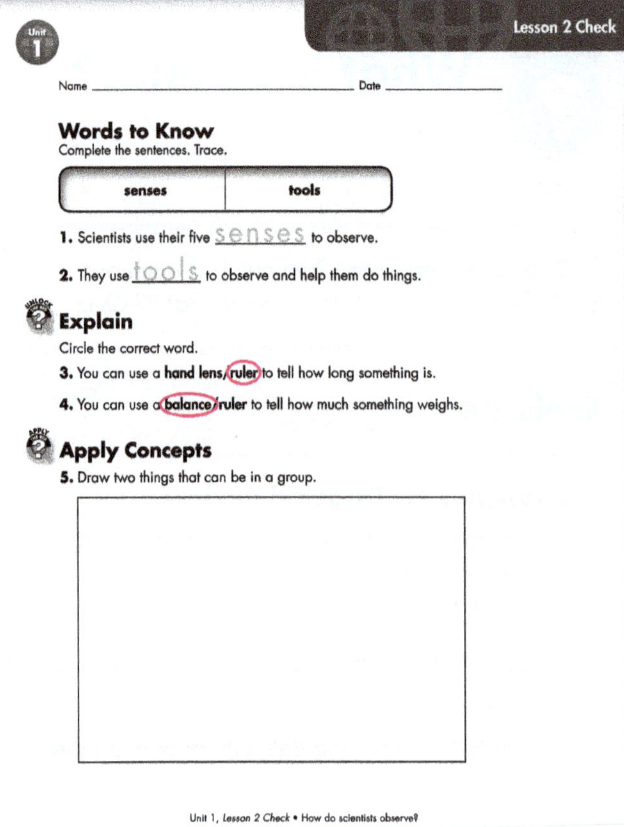

Name _____________________ Date _____________

Words to Know
Complete the sentences. Trace.

senses	tools

1. Scientists use their five <u>senses</u> to observe.

2. They use <u>tools</u> to observe and help them do things.

Explain

Circle the correct word.

3. You can use a **hand lens** / **(ruler)** to tell how long something is.

4. You can use a **(balance)** / **ruler** to tell how much something weighs.

Apply Concepts

5. Draw two things that can be in a group.

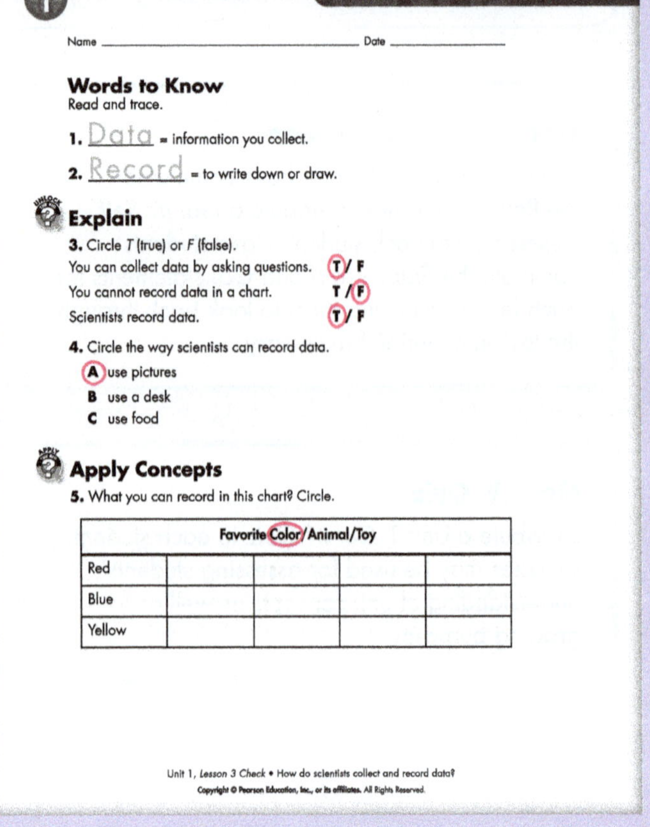

Name _____________________ Date _____________

Words to Know
Read and trace.

1. <u>Data</u> = information you collect.

2. <u>Record</u> = to write down or draw.

Explain

3. Circle *T* (true) or *F* (false).

You can collect data by asking questions. **(T)** / F

You cannot record data in a chart. T / **(F)**

Scientists record data. **(T)** / F

4. Circle the way scientists can record data.

(A) use pictures

B use a desk

C use food

Apply Concepts

5. What you can record in this chart? Circle.

Favorite **(Color)** / Animal / Toy			
Red			
Blue			
Yellow			

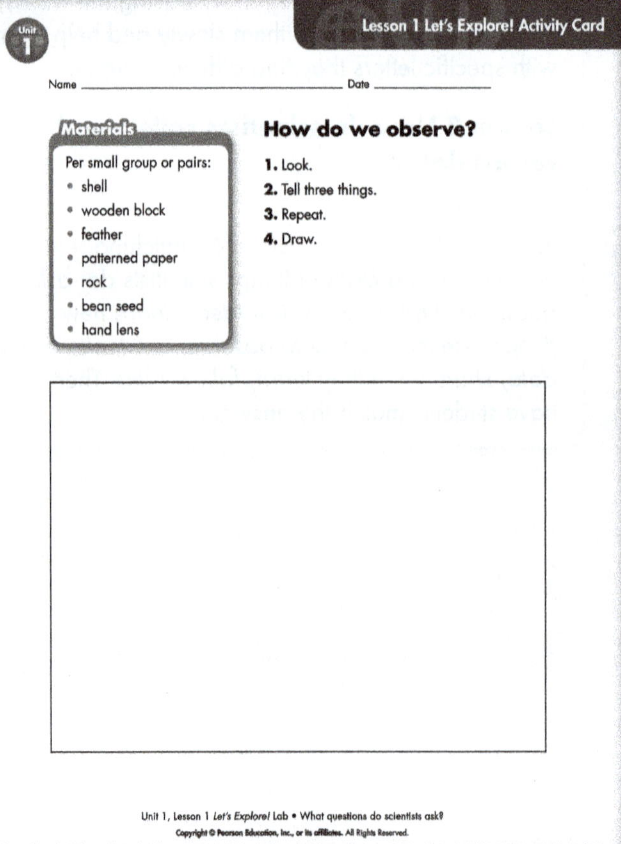

Name _____________________ Date _____________

Materials

Per small group or pairs:
- shell
- wooden block
- feather
- patterned paper
- rock
- bean seed
- hand lens

How do we observe?

1. Look.
2. Tell three things.
3. Repeat.
4. Draw.

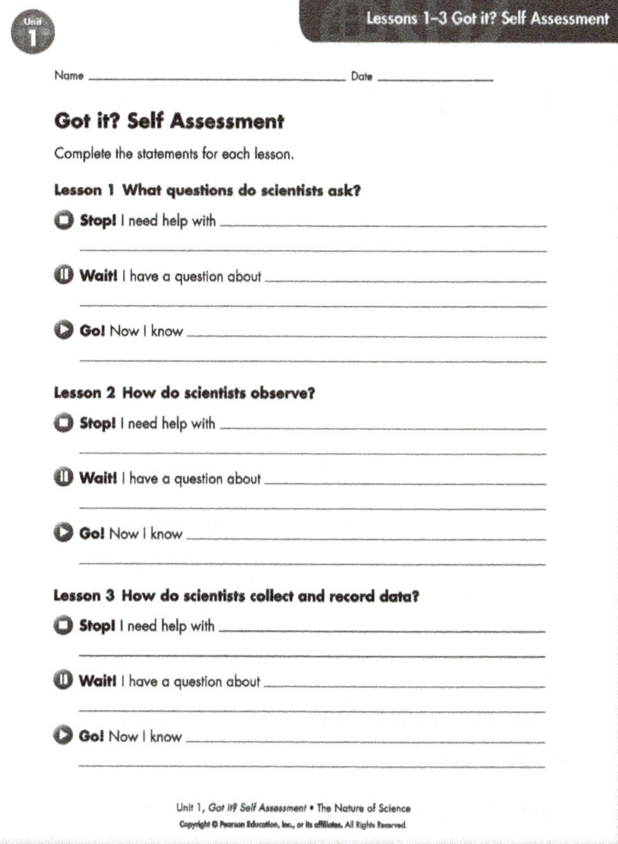

Name _______________ Date _______________

Analyze and Conclude

4. Read and circle.
How does the viewer help you observe things?

It makes things look **smaller**/**bigger**

5. Pick one object you observed. Draw how it looks with water. Draw how it looks without water.

With Water	Without Water

Name _______________ Date _______________

Got it? Self Assessment

Complete the statements for each lesson.

Lesson 1 What questions do scientists ask?

⏹ **Stop!** I need help with _______________
____

⏸ **Wait!** I have a question about _______________

▶ **Go!** Now I know _______________

Lesson 2 How do scientists observe?

⏹ **Stop!** I need help with _______________

⏸ **Wait!** I have a question about _______________

▶ **Go!** Now I know _______________

Lesson 3 How do scientists collect and record data?

⏹ **Stop!** I need help with _______________

⏸ **Wait!** I have a question about _______________

▶ **Go!** Now I know _______________

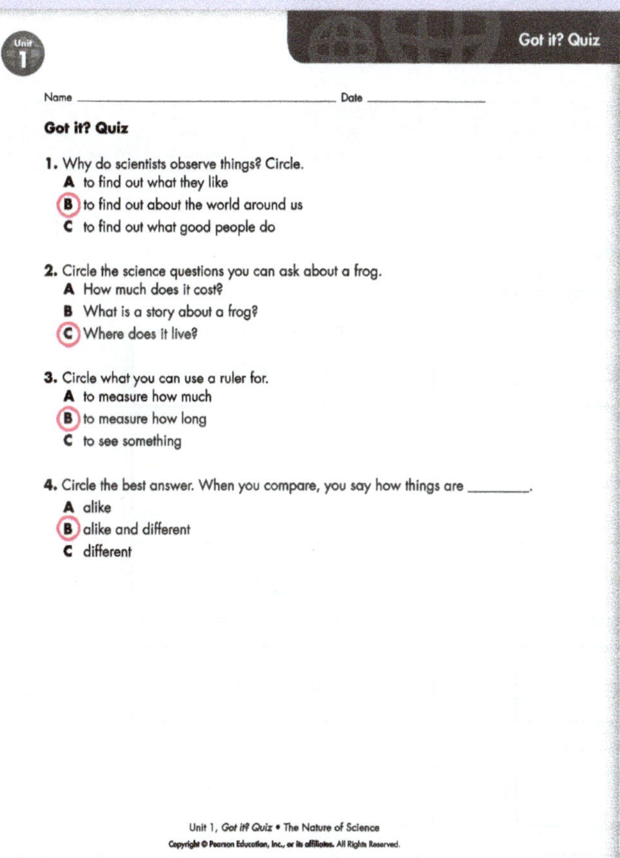

Name _______________ Date _______________

Got it? Quiz

1. Why do scientists observe things? Circle.
  A to find out what they like
  (B) to find out about the world around us
  C to find out what good people do

2. Circle the science questions you can ask about a frog.
  A How much does it cost?
  B What is a story about a frog?
  (C) Where does it live?

3. Circle what you can use a ruler for.
  A to measure how much
  (B) to measure how long
  C to see something

4. Circle the best answer. When you compare, you say how things are _______.
  A alike
  (B) alike and different
  C different

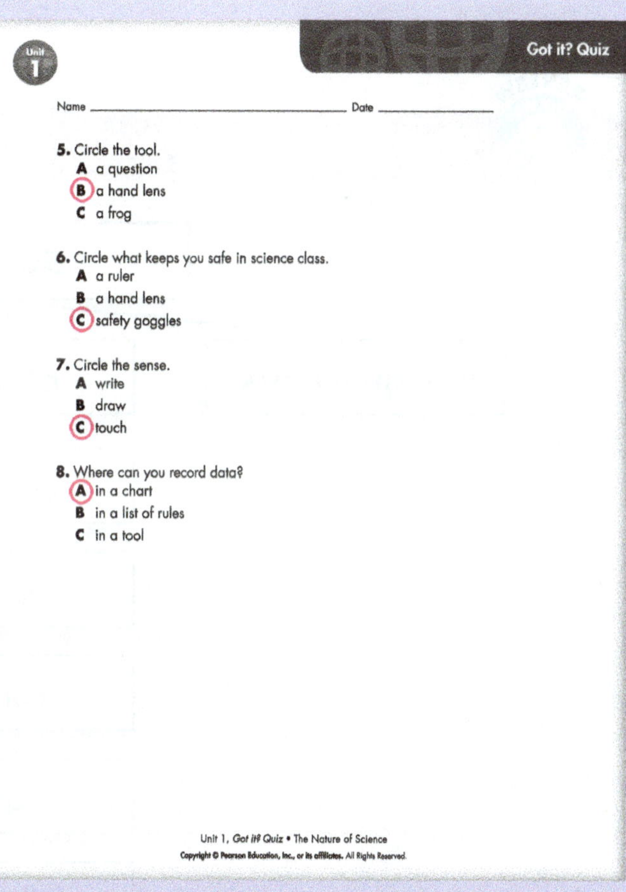

Name _______________ Date _______________

5. Circle the tool.
  A a question
  (B) a hand lens
  C a frog

6. Circle what keeps you safe in science class.
  A a ruler
  B a hand lens
  (C) safety goggles

7. Circle the sense.
  A write
  B draw
  (C) touch

8. Where can you record data?
  (A) in a chart
  B in a list of rules
  C in a tool

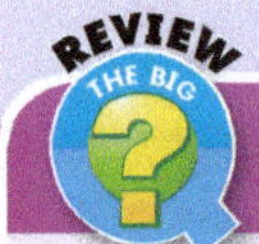

Unit 1 Study Guide

What is science?

Lesson 1
What questions do scientists ask?

- Scientists use science to learn about the world around us.
- Scientists observe and ask questions.

Lesson 2
How do scientists observe?

- Scientists use the five senses and tools.
- They compare things. They group things. They follow rules.

Lesson 3
How do scientists collect and record data?

- Scientists can use words, pictures, numbers, and charts to record data.

Review the Big Question

What is science?

Have students use what they have learned from the unit to answer the question in their own words.

How has your answer to the Big Question changed since the beginning of the unit? What are some things you learned that caused your answer to change?

Make a Concept Map

Draw on the board a concept map like the one shown on this page. With the students, talk through the key ideas from this unit. Invite different students to point to the ideas on the board, miming as possible.

Unit 1 Concept Map

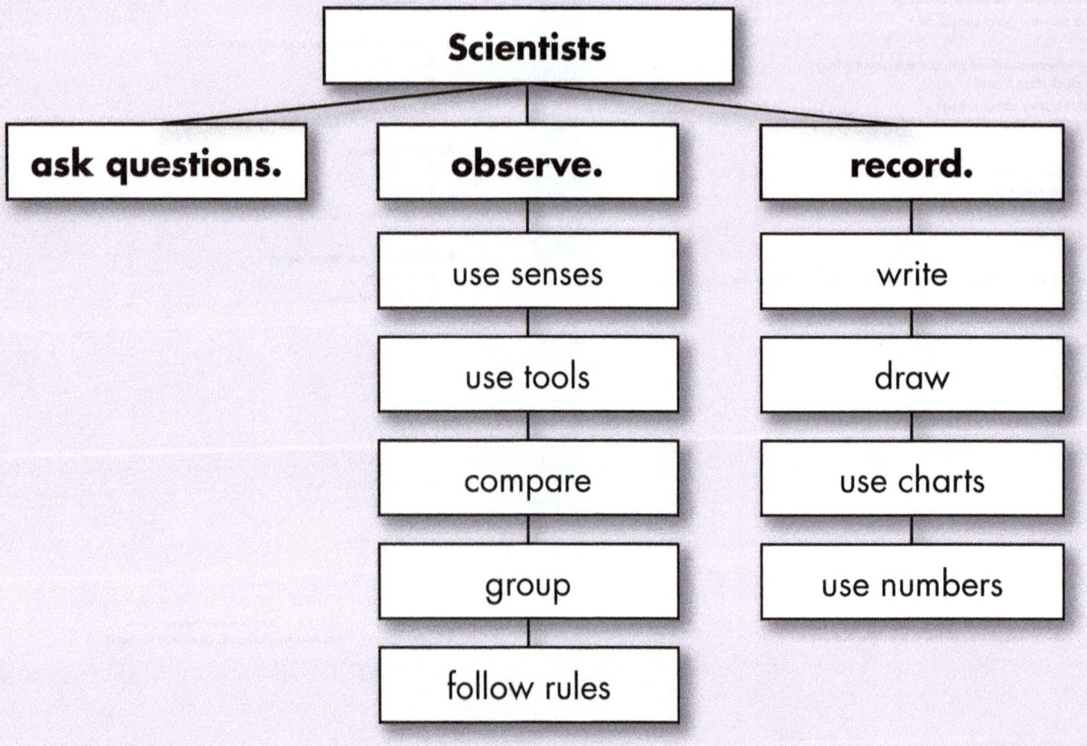

Scientists
- ask questions.
- observe.
  - use senses
  - use tools
  - compare
  - group
  - follow rules
- record.
  - write
  - draw
  - use charts
  - use numbers

Students can make a concept map to help review the Big Question.

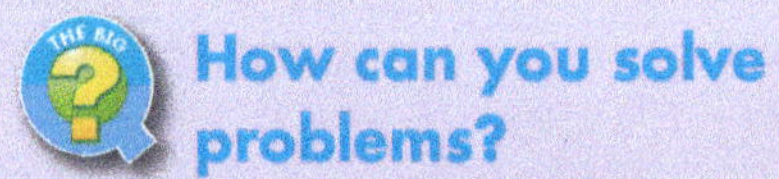

Lesson Plan

Unit Opener & Lesson 1 What are problems and solutions?			
	Activity	**Pages**	**Time**
Engage	• Unit Opener: Think! *Why is the boy wearing eyeglasses?*	SB p. 16	5 min
	• Unit Opener: Identify a problem and solution.	SB p. 16	10 min
	• Unit Opener: Share a solution to solve a problem.	SB p. 16	10 min
Explain	• Problems and solutions	SB p. 17–18	60 min
	• Developing solutions	SB p. 19	30 min
Elaborate	• Solution	TB p. 17	20 min
	• Flash Lab: A Solution at School	SB p. 18	20 min
	• At-Home Lab: Problems and Solutions	SB p. 19	20 min
Evaluate	• *Lesson 1 Check* (ActiveTeach)	TB p. 27a	10 min
	• Assessment for Learning	TB p. 19	10 min
	• Review (Lesson 1)	SB p. 27	10 min
	• *Got it? Self Assessment* (ActiveTeach)	TB p. 27b	10 min
	• *Got it? Quiz* (ActiveTeach)	TB p. 27b	10 min

Lesson 2 How do ideas become solutions?			
	Activity	**Pages**	**Time**
Engage	• Think! *Is it a good design?*	TB p. 21	5 min
	• Think! *Is a pencil holder a tool?*	TB p. 22	5 min
	• Think! *Is paper or plastic a good material for a straw?*	TB p. 23	10 min
Explore	• Digital Lab: *What can this object do?* (ActiveTeach)	TB p. 20	30 min
Explain	• Making plans	SB p. 20	30 min
	• Making designs and choosing tools	SB p. 21–22	60 min
	• Choosing materials	SB p. 23	30 min
Elaborate	• Who makes designs?	TB p. 21	30 min
	• Choose Tools	TB p. 22	10 min
	• Make a Pencil Holder	TB p. 23	30 min
Evaluate	• *Lesson 2 Check* (ActiveTeach)	TB p. 27a	10 min
	• Assessment for Learning	TB p. 23	10 min
	• Review (Lesson 2)	SB p. 27	10 min
	• *Got it? Self Assessment* (ActiveTeach)	TB p. 27b	10 min
	• *Got it? Quiz* (ActiveTeach)	TB p. 27b	10 min

	Activity	Pages	Time
Engage	• Think! *Does the boy's solution work? How do you know?*	SB p. 24	10 min
Explain	• Testing solutions	SB p. 24	30 min
	• How to share solutions	SB p. 25	30 min
	• *Got it? 60-Second Video* (ActiveTeach)	TB p. 25	10 min
Elaborate	• Pencil Holder Test	TB p. 24	20 min
	• Sharing Solutions	TB p. 25	20 min
Evaluate	• *Lesson 3 Check* (ActiveTeach)	TB p. 27a	10 min
	• Assessment for Learning	TB p. 25	10 min
	• Review (Lesson 3)	SB p. 27	10 min
	• *Got it? Self Assessment* (ActiveTeach)	TB p. 27b	10 min
	• *Got it? Quiz* (ActiveTeach)	TB p. 27b	10 min
Lab	• *Let's Investigate! How can you lift heavy things?* (ActiveTeach)	SB p. 26	30 min

Flash Cards

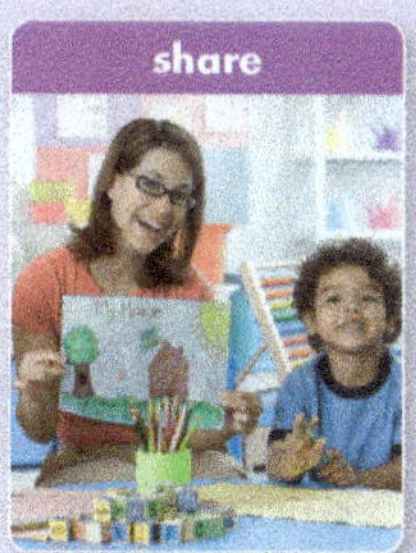

Lesson 1

Key Words	ELL Support
problem, solve, solution	**Vocabulary:** *spill, fix, straw, mud, backpack, boots, umbrella, wrong, barefoot, need, share, eyeglasses/ glasses, lots of, rain, drop, dog, bath, small, big* **Irregular Plurals:** *child/ children, person/people, fish/fish, mouse/mice, tooth/teeth* **Stressed Syllables**

Lesson 2

Key Words	ELL Support
idea, plan, design, choose, materials	**Vocabulary:** *pencil holder, need, pencil, scissors, paper clips, crayon, ruler, balance, forceps/ tweezers, craft sticks, dropper, hand lens, plastic, paintbrush,* **Plurals:** *pencil/pencils, paper clip/paper clips, scissors, goggles, jeans, clothes*

Lesson 3

Key Words	ELL Support
test, change, share, use	**Vocabulary:** *write, talk, draw, show*

Unit 2 — Solve Problems

Unit Objectives

Lesson 1: Students will learn what problems and solutions are.

Lesson 2: Students will learn how ideas become solutions.

Lesson 3: Students will learn how to test and share solutions.

Vocabulary: *wrong, problem, solution, solve, barefoot, need, share, eyeglasses/glasses*

Introduce the Big Question

How can you solve problems?

Build Background Mime trying to write on the board but not having a tool to do so. Ask *What's wrong? I don't have a pen. That's a problem! I can't write on the board without a pen.* Next, mime trying to sit down without a chair. *What's wrong?* Encourage students to explain or mime that you can't sit down because you don't have a chair. *Right! That's a problem! What can solve the problem? A chair! Yes, a chair can be a solution to the problem!*

Engage

Think!

Why is the boy wearing eyeglasses?

Draw students' attention to the picture of the boy at the bottom of the page. Point to the boy's glasses and ask if any students in the class wear glasses, too, or if they know anyone who does. Read the question aloud. Allow students time to discuss the answer. Monitor and provide support as needed.

1 What is wrong? What is the problem? Say with a partner.

Draw students' attention to the picture of the child's feet and pre-teach *barefoot*. *What do you think barefoot means? The child doesn't have shoes!* Put students into pairs and ask *Is it a problem that the child doesn't have shoes?* Have pairs discuss.

ELL Vocabulary Support

Write *foot* on the board and elicit the irregular plural *feet*. Elicit from the class other irregular plural nouns they know, e.g., *child/children, person/people, fish/fish, mouse/mice, tooth/teeth,* etc. Write them on the board and practice the pronunciation with the class.

Unit 2 — Solve Problems

How can you solve problems?

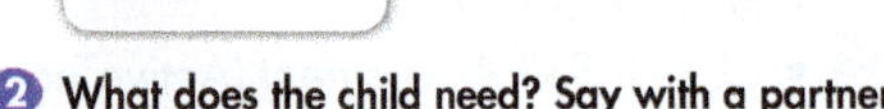

1 What is wrong? What is the problem? Say with a partner.

2 What does the child need? Say with a partner.

3 Why is the girl giving her pen to her classmate? Say as a class.

16   Unit 2

2 What does the child need? Say with a partner.

Read the question aloud and have the same pairs discuss. Invite pairs to share their answers with the class. *The child needs shoes. Why do you need to wear shoes? So you don't hurt your feet and so that you keep them warm!*

3 Why is the girl giving her pen to her classmate? Say as a class.

Focus students' attention on the girl at the bottom of the page. *What is she doing? Why is she giving her pen to her classmate?* Guide students to answer that her classmate doesn't have a (green) pen. *Right! The girl helps her friend solve a problem. She shares her pen!*

Think! Again!

Revisit the question *Why is the boy wearing eyeglasses?*

Invite students to share their ideas freely. (Possible answers: *The boy needs glasses to look at the board in class. The boy needs glasses to read and use a computer or tablet.*) Provide vocabulary support as needed and accept all logical answers.

What are problems and solutions?

> **Objective:** Learn what problems and solutions are.
>
> **Vocabulary:** *spill, problem, wrong, solve, fix, solution, straw, mess, pick up*
>
> **Digital Resources:** Flash Cards (*problem, solve, solution*) *I Will Know...* Digital Activity
>
> **Materials:** glass of water, paper towels, straw

Unlock the Big Question

Write the following text on the board: *How can you solve problems?* Ask *What can solve the problem for the barefoot child? Shoes or socks!*

Build Background Clear off a table and put the glass of water on it. Have the paper towels and straw nearby. Pick up the glass to drink, but spill a little water on the table. *Oh, no! I spill the water when I drink! That's a problem because my books might get wet! How can I solve it?* Show the paper towels. *Do paper towels soak up water? Yes! I can use paper towels to clean up the mess. Paper towels can be the solution!*

Leave the materials out for the next exercises.

Explain

1 Read. Underline what a problem is.

Point to the picture of the juice and invite students to say or mime what the problem is. Read the first paragraph aloud for students and have students underline.

2 Read. Underline what a solution is.

Invite students to read silently and underline what a solution is. Check answers to exercises 1 and 2 as a class.

Display the Flash Cards and discuss each one with students to reinforce understanding.

ELL Vocabulary Support

Write the lesson vocabulary on the board and underline the stressed syllables. Pronounce the words for students and have them repeat.

Lesson 1 · What are problems and solutions?

Key Words
- problem
- solve
- solution

1 Read. Underline what a problem is.

Problems

You can spill when you drink from a glass or cup. This is a problem. A **problem** means there is something wrong. You need to find a way to fix it, or **solve** it.

2 Read. Underline what a solution is.

Solutions

A **solution** is an answer to a problem. What can solve the problem with the juice? A straw can solve the problem. A straw can be the solution.

3 What is the solution to the child's problem? Say with a partner.

I Will Know...   Unit 2   **17**

---

3 What is the solution to the child's problem? Say with a partner.

Have pairs share their answers with another pair. Then say *I'm thirsty. But I can spill when I drink. That's a problem! What can solve it?* Show students the straw and elicit or say *A straw! Right! I can solve the problem. I can use a straw to drink! What is the solution to the problem? A straw!*

Point to the picture of the boy drinking from a straw, read the question aloud, and allow students to discuss freely. Guide them to explain how the straw helps to prevent spills. *Right! You don't have to pick up the glass!*

Elaborate

Solution

Challenge students to say or draw other problems and solutions. (Possible answers: *dropping the glass/straw, not having anything to write with/pencil or notebook, being too cold/wearing a sweater, etc.*)

> **I Will Know...**
>
> Have students do the *I Will Know...* Digital Activity.

What are problems and solutions?

Objective: Think about problems and how to solve them.

Vocabulary: *mud, lots of, rain, backpack, boots, umbrella, drop*

Digital Resources: Flash Cards (*problem, solve, solution*)

Build Background *When you come to school, do you have a lot of things to carry? How do you solve the problem?* Use a backpack! Display the Flash Cards and elicit the words. Encourage students to think of a problem they had recently and to say or mime how they solved it.

Explain

4   Look at the pictures of problems. What are the problems? Say as a class.

Draw students' attention to the first row of pictures. Read or have volunteers read the labels. Make a two-column chart on the board labeled *Problems* and *Solutions*.

As a class, identify the problems and write some of them on the board. (Possible answers: *It is muddy. Her boots are dirty. The boy has a lot of books. He can drop the books. It's raining. The girl gets wet.*)

5   Look at the pictures of solutions. Match the solutions to the problems.

Have students look at the second row of pictures and identify the items by reading the labels. Invite volunteers to say which problem each one solves from exercise 4. Record them in the appropriate columns on the board. Finally, have students match the solutions to the problems in their books.

6   What do you think of the solutions? Talk as a class.

Divide the class into small groups and ask them to discuss how the solutions solve the problems. Elicit ideas from the class.

Ask groups to think of one or two more solutions for each problem in exercise 4 and decide which is the best one. (Possible answers: mud: *sponge to clean the boots, hairdryer to dry the pants;* lots of books: *small suitcase, box;* rain: *raincoat, hat*) Monitor and provide support as necessary. Invite volunteers from each group to explain their best solution for each problem.

---

4 Look at the pictures of problems. What are the problems? Say as a class.

5 Look at the pictures of solutions. Match the solutions to the problems.

6 What do you think of the solutions? Talk as a class.

Flash Lab

A Solution at School
As a class, talk about a problem at school. Talk about a solution.

Elaborate

⚡ Flash Lab

A Solution at School

As a class, brainstorm possible problems children have at school. Write them on the board.

Assign a problem to each group and ask it to discuss possible solutions. Encourage students to list the solutions from most effective to least effective. Students may draw their lists. Then invite groups to share their ideas and discuss as a class.

What are problems and solutions?

> **Objective:** Discuss more problems and solutions.
>
> **Vocabulary:** *dog, bath, small, big*
>
> **Digital Resources:** Flash Cards (*problem, solution*), *Lesson 1 Check* (print out 1 per student)

Build Background Display the Flash Cards side by side. Under the *problem* Flash Card, draw a car with smoke coming out of the front. Explain that your car has broken down, and it's a big problem for you. Elicit solutions for this problem and write them on the board under the *solution* Flash Card. (Possible answers: *take it to the mechanic, take a bus* or *train, ride your bike to school, walk*) Discuss with the class which is the best or fastest solution and why.

Explain

7 **Look at the picture. Is it a problem or a solution? Say as a class.**

Draw students' attention to the picture of the dog. Read the question out loud and say answers as a class. (*The dog is dirty. A bath is a solution. The tub is too small for the dog. It's a problem.*)

8 **Look at the picture. Underline the boy's problem.**

Draw students' attention to the illustration. Invite a volunteer to describe what the boy is doing. (Answer: *He is looking for something.*) Have students underline the problem and check answers. *What is the problem? Right! The boy can't find the green colored pencil! Why can't he find it?* It is on the ground behind the desk where he can't see it.

9 **What is a solution? Say as a class.**

Read the question aloud and invite students to brainstorm possible solutions. If necessary, draw their attention to the picture of the pencil holder. Invite volunteers to explain how a pencil holder can be a solution to the boy's problem.

---

7 Look at the picture. Is it a problem or a solution? Say as a class.

8 Look at the picture. Underline the boy's problem.

a) The boy can't find the ruler.

b) The boy can't find the green colored pencil.

c) The boy can't find the notebook.

9 What is the solution? Say as a class.

At-Home Lab

Problems and Solutions

Look at home for a problem and solution.

Say the problem. Say the solution.

Tell your family.

Lesson 1 Check Unit 2 **19**

Elaborate

At-Home Lab

Problems and Solutions

Assign the lab as homework. Ask students to write down the problem and solution they identify and report to the class the next day.

Evaluate

Lesson 1 Check **Assessment for Learning**

Review the Key Words for Lesson 1 (see Student's Book page 17). Distribute the *Lesson 1 Check* and guide students as they complete it. Check answers as a class. Then ask students to grade their progress on the topic of problems and solutions from 1 to 3: 3 = *I understand what problems and solutions are;* 2 = *I need to study more;* 1 = *I need help!* Encourage students giving themselves a 1 or a 2 to say what they found difficult and what they need to study more.

Lesson 2

How do ideas become solutions?

Objective: Learn how to make a plan.

Vocabulary: *problem, idea, solution, plan, forceps/tweezers, craft sticks, dropper, paper clips, hand lens*

Digital Resources: Flash Cards (*tools, idea*), *Let's Explore!* Digital Lab

Materials: forceps/tweezers, craft sticks, plastic dropper, paper clips, hand lens, two pieces of paper, small beads or edible seeds, a cup with water

Unlock the Big Question

Write the following on the board: *How do ideas become solutions?* Invite volunteers to present a "home" problem they detected in the previous lesson and the ideas they had to solve it. Point out how ideas can become solutions.

Build Background *We've talked about some different problems and some solutions. Sometimes, you can't solve a problem right away. Sometimes, you have to have an idea for a solution and make it.* Explain to students that they will learn about how to turn ideas into solutions.

Explore

Let's Explore! Lab What can this object do?

Objective: Learn about more tools and what they do.

Digital Resources: *Let's Explore!* Digital Lab, *Let's Explore!* Activity Card (1 per student)

- Display the *tools* Flash Card. Elicit from students what a tool is. (*Something that helps you do something. Some tools help you observe things.*) Elicit some examples students already know and what they can help you do. (Possible answers: *hand lens/see, ruler/measure, pencil/draw*) Explain that students will learn about some more tools in this lab.

- Display the dropper. *What does this tool do? It measures! Right! You can measure drop by drop.* Allow students to examine the remaining tools and think about what they might be used for. Write the names on the board: *dropper, forceps, craft sticks, paper clip.*

- Provide the rest of the materials and allow students to experiment with them and show how you can use them. *Right! You can pick things up with forceps.*

- Show the Digital Lab. Have students work in pairs to complete the *Activity Card.* Monitor and provide support as needed.

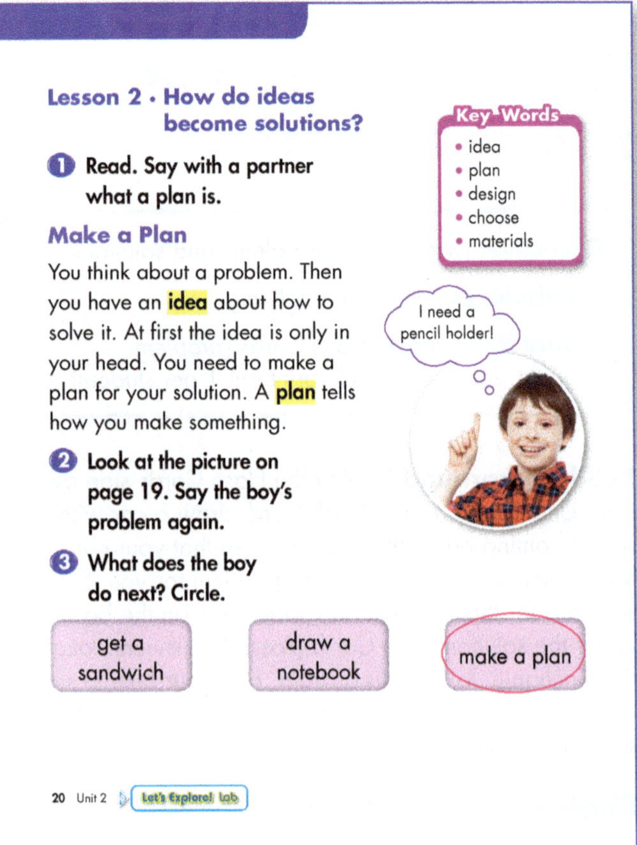

Explain

1 **Read. Say with a partner what a plan is.**

Read the paragraph for students. Have pairs review the paragraph and say what a plan is. Activate prior knowledge by having students brainstorm some instances in which they have made plans.

Have you or your family ever made a plan to do something? (Possible answers: *Pack a lunch. Do my homework. Go on a trip.*) *Have you ever made a plan to make something? Did you draw the plan?* Accept all logical answers.

ELL Reading Strategy Support

You may wish to point out the Key Words box and that the key words *idea* and *plan* are highlighted in yellow in the text. Point out that students can also highlight words to help remember them.

2 **Look at the picture on page 19. Say the boy's problem again.**

Invite the class to say the boy's problem. *He can't find his pencil!* Point to the speech bubble and read it or have a volunteer do so. Then read the instruction aloud and have students say in pairs what the boy's idea for a solution is. (Possible answers: *Get a pencil holder. Make a pencil holder.*)

3 **What does the boy do next? Circle.**

The boy wants to make a pencil holder? What does he do next? Invite students to circle the answer.

How do ideas become solutions?

> **Objective:** Learn about making a design.
>
> **Vocabulary:** *draw, design, solution*
>
> **Digital Resources:** Flash Card (*design*), *I Will Know…* Digital Activity

Build Background Display the *design* Flash Card and invite students to say what the boy is doing. Draw a pencil holder on the board. *What's this? Who needs it? Why? How can he plan to make a pencil holder?* Elicit ideas from the class.

Explain

4 Read. Say how you can show your plan.

Read the paragraph along with students. Activate prior knowledge by discussing any designs students have seen or made.

5 Look at the picture. What does the boy draw? Circle.

Have students do the exercise. Invite volunteers to share their answers. *What is the boy's design? A pencil holder! Right. His drawing shows what it will look like in the end.*

6 Draw your own design for a pencil holder.

Ask the students to spend a few minutes thinking about their own pencil holders. *How big does it need to be? What shape do you want it to be?* Encourage students to close their eyes and visualize it. Then give them time to make designs for their pencil holders. Suggest that they make a few sketches first on a piece of paper and then draw the final design in their books. Monitor and provide help as necessary.

Invite students to show their designs in small groups or to the class.

Elaborate

Who makes designs?

Brainstorm with the students jobs in which people need to make designs. Write them on the board and discuss what types of designs these are. (Possible answers: *for a house, for clothes, for a bridge,* etc.)

Ask the students to think about each job and decide which one they would like to do and why. Invite volunteers to explain.

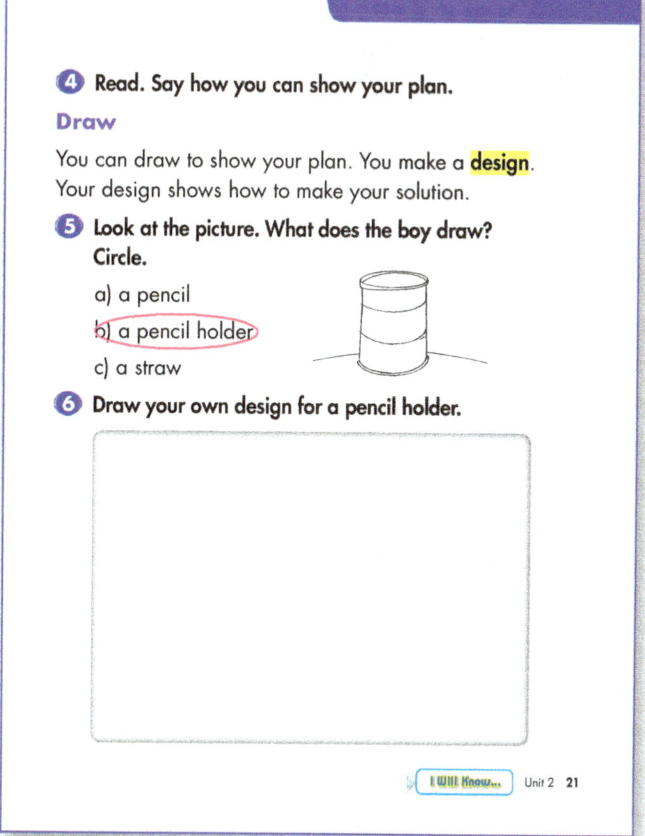

Think!

Is it a good design?

Ask the students to look at their backpacks and think about the design. *Are you happy with the design of your backpack? Are there any problems with it?* Invite volunteers to talk about their backpacks and whether they are happy or not with the design.

> **I Will Know...**
>
> Have students do the *I Will Know…* Digital Activity.

How do ideas become solutions?

> **Objective:** Learn how to choose tools.
>
> **Vocabulary:** *choose, tools, pencil, scissors, paper clips, crayon, ruler, balance*
>
> **Digital Resources:** Flash Card (*choose*)
>
> **Materials:** different kinds of tools (for example, crayon, pencil, pen, ruler, scissors, stapler, paper clip, eraser)

Build Background Display the *choose* Flash Card. Elicit or describe for students what's on each of the plates. *The girl has to choose. She has to say which thing she wants to eat! What do you think she will choose? The fruit or the cookies and candy?* Explain that *choose* means *to pick* or *to decide.*

Explain

7 **Read and look. Trace.**

Review the two steps students have seen about how to come up with solutions. (Possible answers: *Think of an idea. Make a design.*) Then read the first line and write *Tools* on the board.

Activate prior knowledge by having students name all the tools they can. Have them look at the pictures and identify the tools. Read aloud what each tool can do. For each one, ask whether students agree and if they have anything more to add.

Give students time to trace the words. Then say the words for them to repeat.

ELL Language Support

Draw two columns on the board with headings: *One* and *Two+. We say one pencil. Two?* Elicit *pencils.* Write *pencil* and *pencils* under the corresponding headings on the board. Repeat with *paper clip(s).* Then point out that we don't say *one scissor. Scissors is always plural!* Elicit other words that are used only as plurals, e.g., *glasses, goggles, clothes, jeans,* etc.

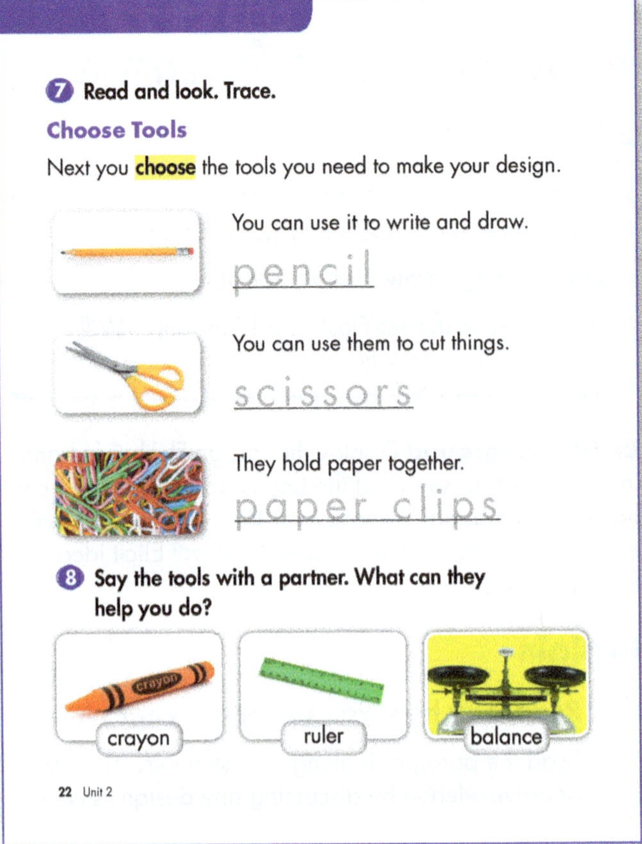

8 **Say the tools with a partner. What can they help you do?**

Have students work in pairs. Read the question aloud for students and have the pairs say the tools and think of what they use them for. Invite pairs to share their answers with the class. Provide support as necessary.

Elaborate

Choose Tools

Place the tools on a table where students can see them. Have pairs or small groups look at the tools and think which they would choose if they were to make their pencil holders. Encourage students to discuss how they would use them. Monitor and provide help as necessary.

Think!

Is a pencil holder a tool?

Ask the question and invite students to discuss freely. Accept all logical answers.

How do ideas become solutions?

> **Objective:** Learn how to choose materials.
>
> **Vocabulary:** *choose, materials, plastic, paintbrush, stickers, can, cotton, wood, fabric*
>
> **Digital Resources:** Flash Cards (*choose, materials*), *Lesson 2 Check* (print out 1 per student)
>
> **Materials:** different kinds of materials (for example, fabric, wood, plastic), empty toilet paper rolls (1 per student), glue, watercolor paint, paintbrushes (1 per student), glitter or sequins of different colors

Build Background Display the materials at the front of the class along with the *materials* Flash Card. *We learned that you need to choose tools to make something. You also need to choose the materials.*

Display the *materials* Flash Card and elicit from students what each kind of material might be good for. (Possible answers: *clothes, furniture*) Provide support as needed.

Explain

9 Read. What are materials? Say with a partner.

Read the paragraph aloud for students. Have pairs say what the materials are and say or point to some examples. Invite pairs to share their answers with the class. Provide support as necessary.

10 Look at the pictures. Circle the materials the boy chooses to make the pencil holder on page 24.

Have students look at the picture of the pencil holder and then at the pictures of the materials. Invite students to identify the items that they can, and name any they cannot. (Answers: *glue, paper, can, construction paper, paintbrush, scissors, stickers*) Have students circle the materials they think the pencil holder is made of. Check answers as a class. Encourage students to explain why they selected the items they did.

Next, have students point to the tools they think were used to make the pencil holder and explain or mime how they were used. Provide vocabulary support as necessary.

11 Read and say.

Draw students' attention to the diagram. Read the steps aloud and, after each one, prompt students to explain what the boy did. For example, *Identify the problem. The boy loses his pencils. Think of an idea. The boy thinks about making a pencil holder.*

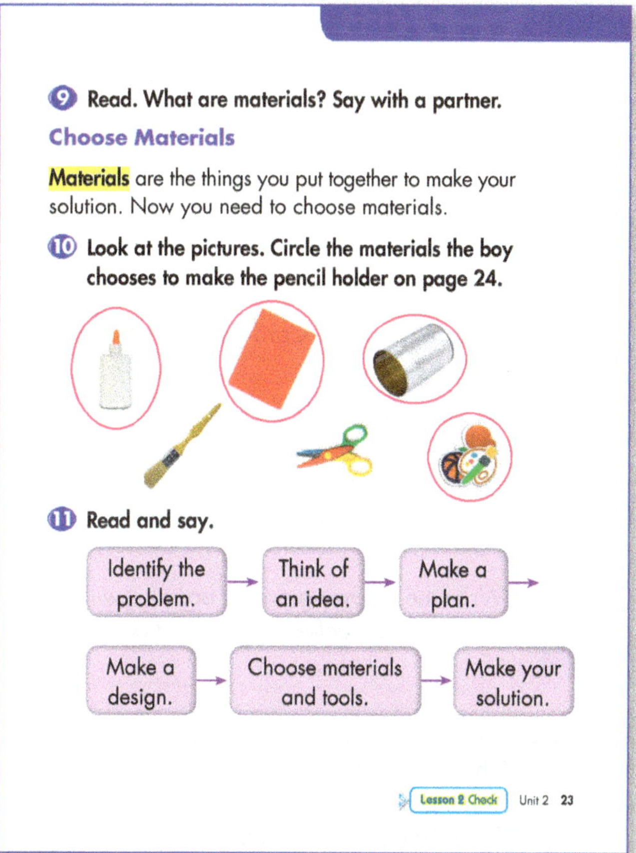

Elaborate

Make a Pencil Holder

Distribute materials and have students make "pencil holders" out of the toilet paper rolls. Have them choose a color or two to paint them and a color of glitter or sequins to glue on as decoration. Have students describe the choices they make and encourage them to say why. (*I like green.*) Save the "pencil holders" for exercises later in the unit.

Think!

Is paper or plastic a good material for a straw?

Remind students that they tested how paper towels soak up water. Invite students to discuss the question freely. (Possible answer: *Paper is not a good material for a straw. Plastic is a good material for a straw.*) They may mime or draw pictures as necessary.

Evaluate

Lesson 2 Check   Assessment for Learning

Review the Key Words for Lesson 2 (see Student's Book page 20). Distribute the *Lesson 2 Check* and guide students as they complete it. Check answers as a class. Then ask students to grade their progress on the topic of *how ideas become solutions*; 2 = *I need to study more*; 1 = *I need help!* Encourage students giving themselves a 1 or a 2 to say what they found difficult and what they need to study more.

How can you test and share solutions?

> **Objective:** Learn how to test and share solutions.
>
> **Vocabulary:** *test, change*
>
> **Digital Resources:** Flash Card (*test*), *I Will Know...* Digital Activity
>
> **Materials:** students' pencil holders

Unlock the Big Question

Write the following on the board: *I will learn how scientists collect and record data.*

Build Background Review the boy's problem and solution and the process of his idea becoming a solution. *What is the boy's problem? He can't find his pencil. What is his idea? To make a pencil holder. What is his design? A pencil holder! Right! His idea can become a solution!* Explain to students that they'll learn what comes next.

> ### ELL Reading Strategy Support
>
> You may wish to take the opportunity to point out the paragraph heading to students. Explain that headings can tell readers what a paragraph is about. *A heading can help you understand what you read.*

Explain

1 **Read. What can the boy do next? Say as a class.**

Read the paragraph for students. Activate prior knowledge by asking them if they have ever tried something to check if it works. Elicit examples and have students say what *test* means in this context (*try out*). Accept all logical answers. Have the class say what the boy does next.

2 **How can you test an umbrella? Say with a partner.**

Focus students' attention on the picture of the umbrella. Elicit or say what it is and what it is used for. Ask the question and have pairs come up with answers. Monitor and provide support as needed. (Possible answers: *Take it out on a rainy day and try it. Put it in the bathtub and throw water over it. Throw a bucket of water over it in the yard.*)

> ### Lesson 3 · How can you test and share solutions?
>
> **1** **Read. What can the boy do next? Say as a class.**
>
> **Key Words**
> - test
> - change
> - share
> - use
>
> #### Test Your Solution
>
> Scientists test ideas. They see if their idea is right. You can **test** your solution. Does your solution solve the problem? If not, you can **change** it. You can change the design or the materials.
>
> **2** **How can you test an umbrella? Say with a partner.**
>
> 
>
> **3** **Look at the picture. How can the boy test his solution? Say as a class.**
>
> 
>
> 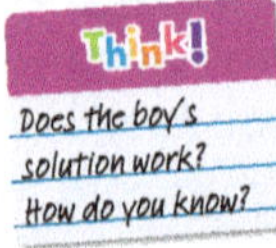
>
>
> 24   Unit 2   I Will Know...

3 **Look at the picture. How can the boy test his solution? Say as a class.**

Display the *test* Flash Card. Have students describe the picture, and ask the question. Invite volunteers to share their answers. *Right! The boy can put pencils in his pencil holder to see if it works. Does it hold pencils? Yes! It works!*

Elaborate

Pencil Holder Test

Distribute the pencil holders and have students test them by putting pencils in them. *Let's test your pencil holders.* Have students discuss or mime why they don't work very well. (Possible answers: *They fall over. They don't have a bottom.*) *Does the pencil holder work? No! Can you use it? No!* Challenge students to come up with suggestions to improve the design. (Possible answer: *We can choose better materials!*)

Think!

Does the boy's solution work? How do you know?

Have students discuss the answer. Invite volunteers to share their answers. (*It holds pencils.*)

> ### I Will Know...
>
> Have students do the *I Will Know...* Digital Activity.

Lesson 3

How can you test and share solutions?

> **Objective:** Learn about sharing solutions.
>
> **Vocabulary:** *share, use* (v)
>
> **Digital Resources:** Flash Card (*share*), *Lesson 3 Check* (print out 1 per student), *Got it? 60-Second Video*

Build Background Activate prior knowledge by recalling what students have learned about scientists and how they work. (Possible answers: *They ask questions, observe, test, collect and record data, etc.*) Talk with students about how turning problems into solutions is similar to what scientists do and explain that people who design things often share their ideas and solutions. *Today we will learn more about sharing solutions.*

Explain

4 **Read. Look at the pictures. Who shares their solution? Mark (✔).**

Show the *share* Flash Card and invite students to describe what they see. Read the paragraph aloud with students. Have students mark the pictures of who they think is sharing their solutions. Invite volunteers to explain why they selected the pictures they did.

Next, invite a volunteer to explain what *use* means. Reinforce understanding of *use* by asking questions. *Does the girl use her bike? What do you use to write?*

5 **Trace some ways you can share.**

Have students trace the words. Invite volunteers to read each of the words. Solicit from other volunteers other ways you can share solutions. (Possible answers: *write, use a computer, etc.*)

5 **Read the sentences. Cover your book. Say the sentences with a partner.**

Invite students to read aloud a sentence each. Encourage them to talk about a time when they shared an idea, solution, or information with others.

Then pair students and have them cover their books and say the sentences together. Monitor and help as necessary.

---

4 Read. Look at the pictures. Who shares their solution? Mark (✔).

Share

Scientists **share** their answers. You can share your solutions. You can write, talk, draw, and show pictures. Then other people can **use** your solutions. They use your solutions to solve problems!

5 Trace some ways you can share.

show   tell   draw

6 Read the sentences. Cover your book. Say the sentences with a partner.

1. You think of an idea.
2. You make a plan and draw your solution.
3. You choose materials and tools.
4. You test your solution to see if it works.
5. You share your solution with others.

Lesson 3 Check   Got It?   60-Second Video   Unit 2   25

Elaborate

Sharing Solutions

Divide the class into small groups. Have one student in each group mime missing something. The other students guess what's missing and provide or mime sharing the solution with their classmate. Alternatively, say a problem and have groups draw a solution. Monitor and provide support as needed. (Possible answers: *I can't see the board. You need glasses. I have a lot of heavy books to take to school. You need a backpack! The paper is torn. You need tape!*)

Evaluate

Lesson 3 Check   Assessment for Learning

Review the Key Words for Lesson 3 (see Student's Book page 23). Distribute the *Lesson 3 Check* and guide students as they complete it. Check answers as a class. Then ask students to grade their progress on the topic of how you can test and share solutions from 1 to 3: 3 = *I understand how to test and share solutions;* 2 = *I need to study more;* 1 = *I need help!* Encourage students giving themselves a 1 or a 2 to say what they found difficult and what they need to study more.

> 
> **Got it?   60-Second Video**
> Play the *Got it? 60-Second Video* to review the unit material.

Let's Investigate!

In this unit, students learn about solving problems. In this lab, they will investigate and test solutions for moving something heavy up a ramp.

Let's Investigate! Lab **How can you lift heavy things?**

Objective: Students will learn that they can use science and technology to solve real-world problems and that it is helpful to work in groups to share ideas.

Materials: one set per group: books, craft stick, cardboard, string

Digital Resources: *Let's Investigate!* Digital Lab, *Let's Investigate! Activity Card* (1 per student)

Advance Preparation: Set up a stack of books and place one book in front of the stack. Place the string and craft stick nearby.

- *I want to move the book to the top of the stack, but it's very heavy. How can I solve this problem?*

- Have students look at the materials and brainstorm solutions. Have them refer to the *Activity Card* as necessary.

- Lead students to understand that you can pull the book up the ramp with the string or push it up the ramp with the craft stick. Both solutions make it easier to move the book to the top of the stack than lifting it.

- Have students complete the *Activity Card*. Guide students to understand that working together and sharing ideas and answers can help us solve problems.

Teacher Time-Saving Option: Show the *Let's Investigate!* Digital Lab as an alternative to the hands-on lab activity.

Unlock the Big Question

Have students refer to the Big Question on the Unit Opener page. In pairs have them recall what they have learned about how scientists design and test solutions to problems. Have pairs complete questions 6 and 7 on the *Activity Card*.

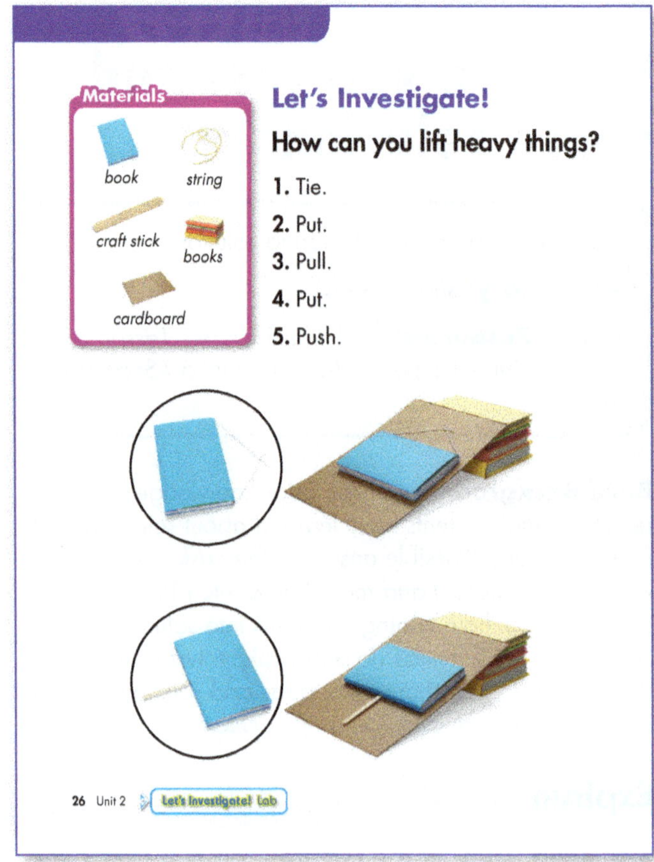

Class Project: Problems and Solutions Poster

Materials: art supplies, cardboard for poster, sheets of A4 paper cut in half

Write the heading in big letters on the poster: *Problems and Solutions*. Show it to the class and explain that students are going to create a poster for the class. Divide the class into eight groups, one for each step of the poster. Refer students to page 23 of the Student's Book and assign one step from exercise 11 to each group. Elicit from the class what the two last, missing steps are. (*Test your solution. Share your solution.*)

Hand out the pieces of paper and ask each group to copy its step on its piece of paper and decorate it. When they finish, gather the pieces of paper and attach them to the poster. You can draw arrows or ask students to do this. Allow students to add any further decoration, and display the poster in class. Invite volunteers to say which is their favorite step and why.

Unit 2 Review

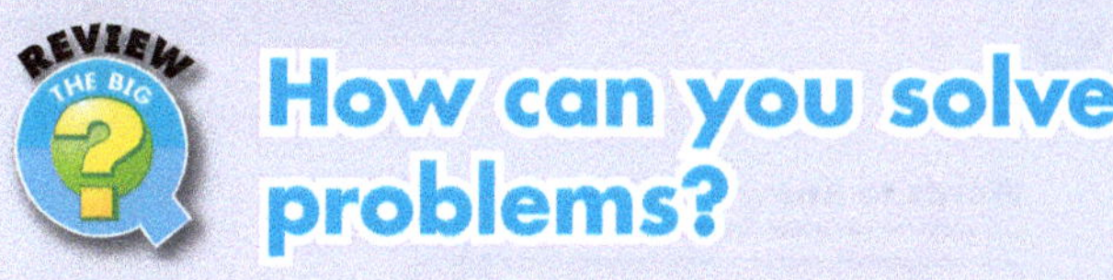

How can you solve problems?

Digital Resources: Print out 1 of each per student: *Got it? Self Assessment, Got it? Quiz*

Evaluate

Strategies for Targeted Review

The following are strategies for providing targeted review for students if they encounter challenges with the content.

Lesson 1 What are problems and solutions?

Question 1

If... students are having difficulty identifying the problem, then... mime spilling a drink and then using a straw.

Lesson 2 How do ideas become solutions?

Question 2

If... students are having difficulty choosing the correct answers, then... review the steps of making a plan, designing your solution, choosing tools, and choosing materials.

Lesson 3 How can you test and share solutions?

Question 3

If... students are having difficulty selecting who is not sharing, then... describe each picture as a class.

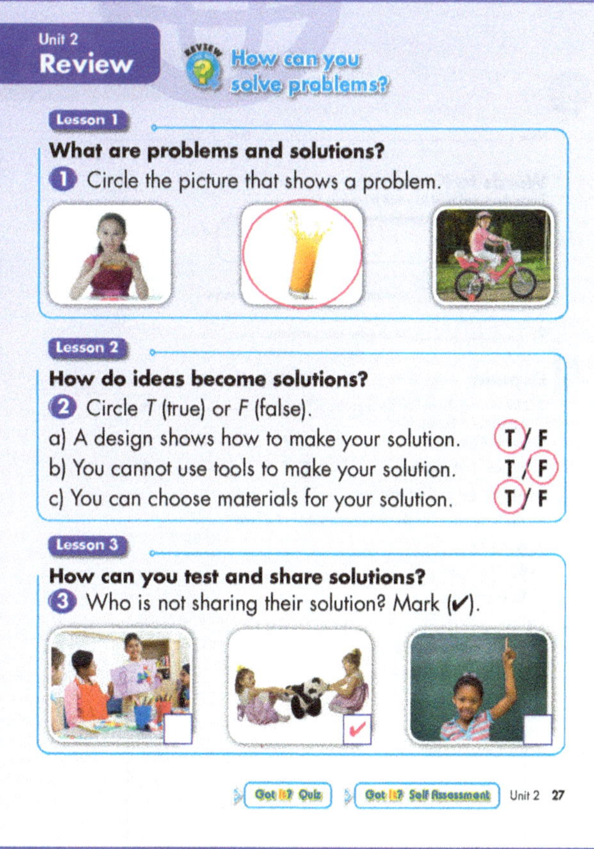

ELL Language Support

Before students start working on the Review activities, have them read each question aloud along with you.

Got it? Self Assessment

Immediately after students have completed the Review activities, distribute a *Got it? Self Assessment* to each student. Have students complete the *Stop! Wait!* and *Go!* statements for each lesson, allowing them to look back through the lesson material if necessary.

Got it? Quiz

Distribute a Unit 2 *Got it? Quiz* to each student. Quizzes may be used for assessing students' understanding of unit concepts as well as for grading purposes.

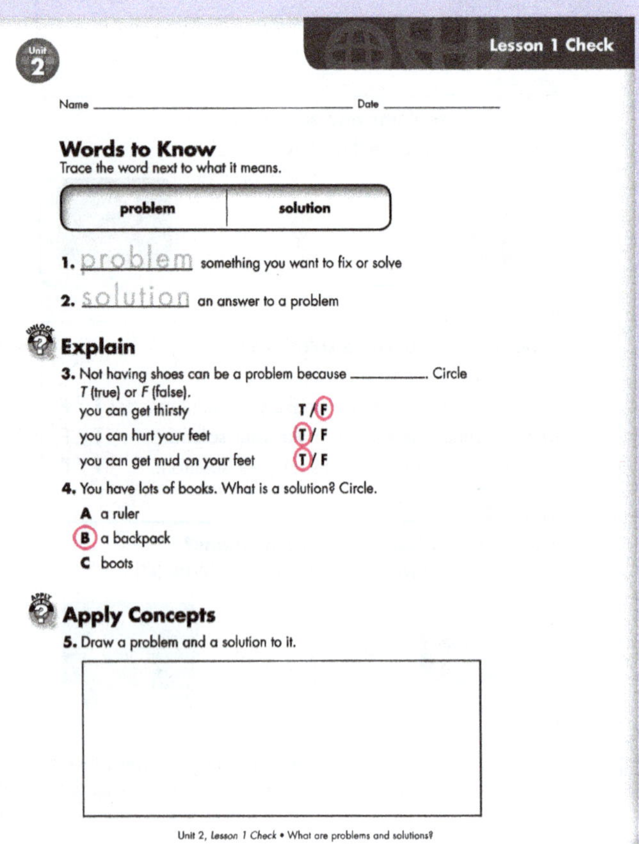

Lesson 1 Check

Name _______________________ Date _______________

Words to Know
Trace the word next to what it means.

problem	solution

1. problem something you want to fix or solve

2. solution an answer to a problem

Explain

3. Not having shoes can be a problem because __________. Circle
T (true) or F (false).
you can get thirsty   T / **F**
you can hurt your feet   **T** / F
you can get mud on your feet   **T** / F

4. You have lots of books. What is a solution? Circle.
A a ruler
B a backpack
C boots

Apply Concepts

5. Draw a problem and a solution to it.

Unit 2, *Lesson 1 Check* • What are problems and solutions?
Copyright © Pearson Education, Inc., or its affiliates. All Rights Reserved.

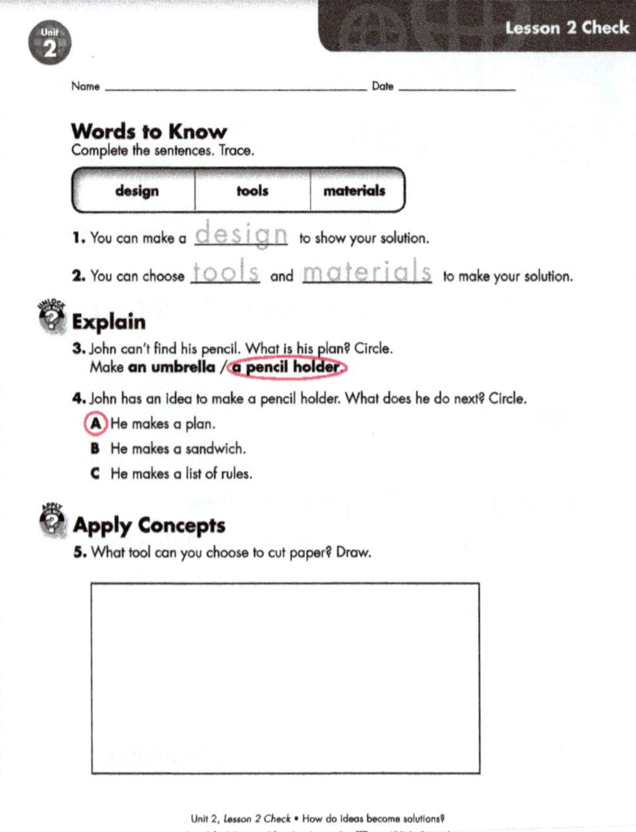

Lesson 2 Check

Name _______________________ Date _______________

Words to Know
Complete the sentences. Trace.

design	tools	materials

1. You can make a design to show your solution.

2. You can choose tools and materials to make your solution.

Explain

3. John can't find his pencil. What is his plan? Circle.
Make **an umbrella** / **a pencil holder**

4. John has an idea to make a pencil holder. What does he do next? Circle.
A He makes a plan.
B He makes a sandwich.
C He makes a list of rules.

Apply Concepts

5. What tool can you choose to cut paper? Draw.

Unit 2, *Lesson 2 Check* • How do ideas become solutions?
Copyright © Pearson Education, Inc., or its affiliates. All Rights Reserved.

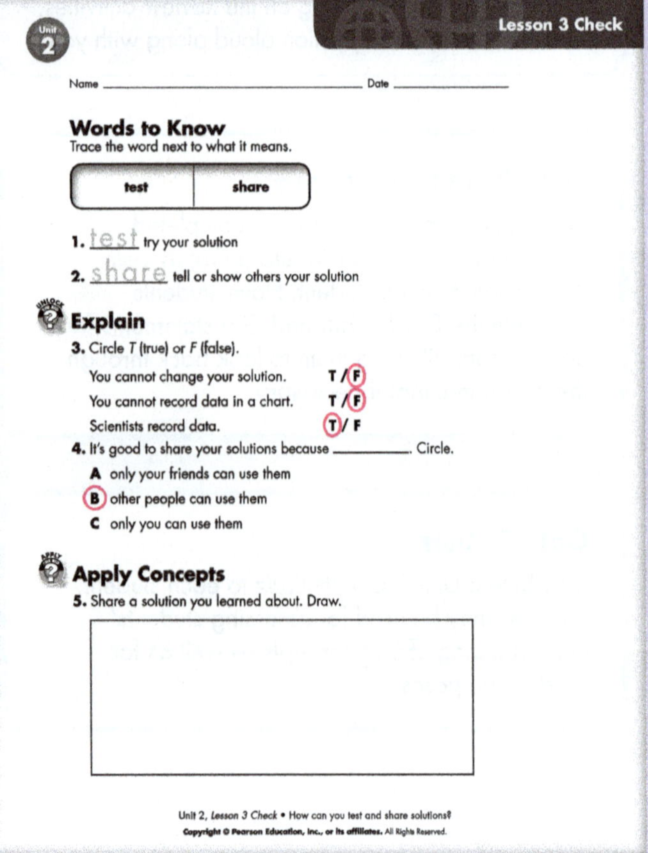

Lesson 3 Check

Name _______________________ Date _______________

Words to Know
Trace the word next to what it means.

test	share

1. test try your solution

2. share tell or show others your solution

Explain

3. Circle T (true) or F (false).
You cannot change your solution.   T / **F**
You cannot record data in a chart.   T / **F**
Scientists record data.   **T** / F

4. It's good to share your solutions because __________. Circle.
A only your friends can use them
B other people can use them
C only you can use them

Apply Concepts

5. Share a solution you learned about. Draw.

Unit 2, *Lesson 3 Check* • How can you test and share solutions?
Copyright © Pearson Education, Inc., or its affiliates. All Rights Reserved.

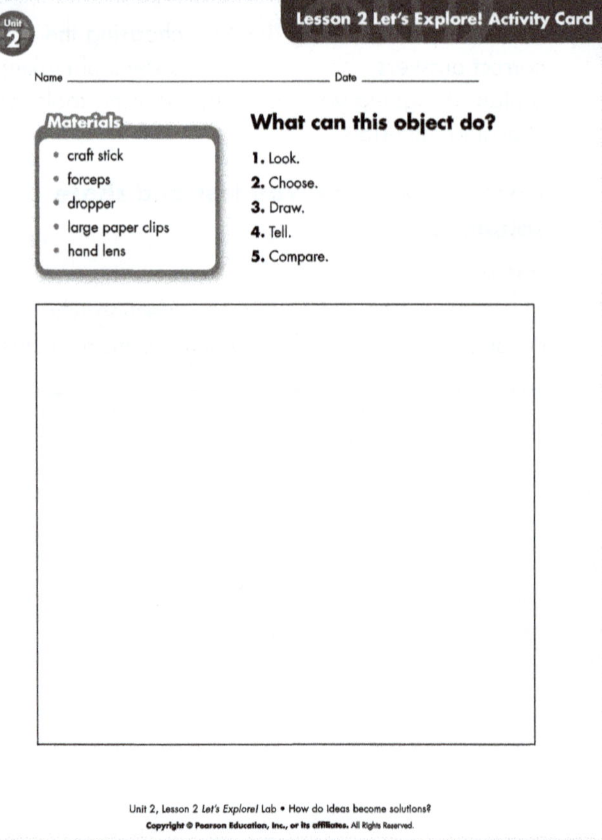

Lesson 2 Let's Explore! Activity Card

Name _______________________ Date _______________

Materials
- craft stick
- forceps
- dropper
- large paper clips
- hand lens

What can this object do?

1. Look.
2. Choose.
3. Draw.
4. Tell.
5. Compare.

Unit 2, *Lesson 2 Let's Explore! Lab* • How do ideas become solutions?
Copyright © Pearson Education, Inc., or its affiliates. All Rights Reserved.

T27a   Unit 2 • Digital Resources and Photocopiables

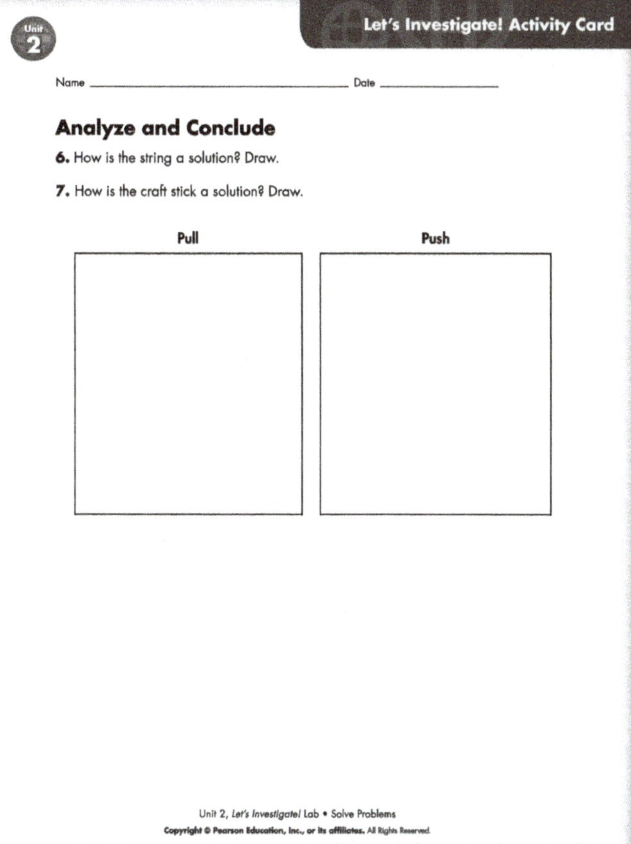

Analyze and Conclude

6. How is the string a solution? Draw.

7. How is the craft stick a solution? Draw.

Pull	Push

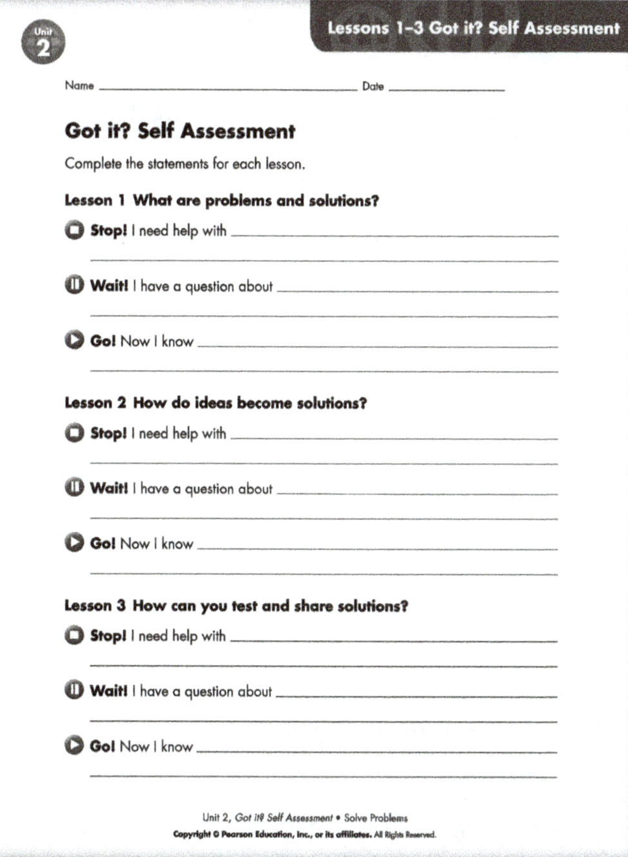

Got it? Self Assessment

Complete the statements for each lesson.

Lesson 1 What are problems and solutions?

Stop! I need help with ______________________

Wait! I have a question about ______________________

Go! Now I know ______________________

Lesson 2 How do ideas become solutions?

Stop! I need help with ______________________

Wait! I have a question about ______________________

Go! Now I know ______________________

Lesson 3 How can you test and share solutions?

Stop! I need help with ______________________

Wait! I have a question about ______________________

Go! Now I know ______________________

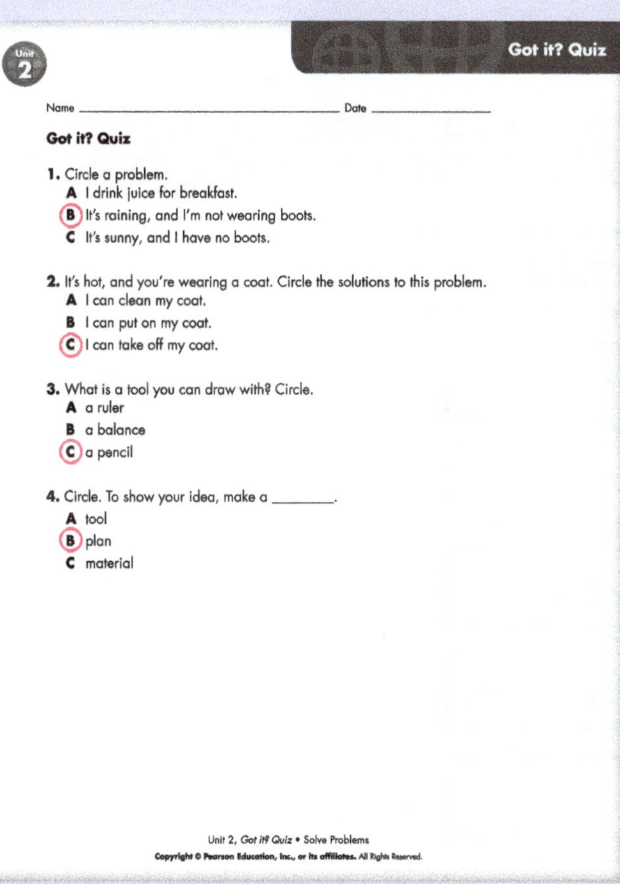

Got it? Quiz

1. Circle a problem.
 A I drink juice for breakfast.
 B It's raining, and I'm not wearing boots.
 C It's sunny, and I have no boots.

2. It's hot, and you're wearing a coat. Circle the solutions to this problem.
 A I can clean my coat.
 B I can put on my coat.
 C I can take off my coat.

3. What is a tool you can draw with? Circle.
 A a ruler
 B a balance
 C a pencil

4. Circle. To show your idea, make a ________.
 A tool
 B plan
 C material

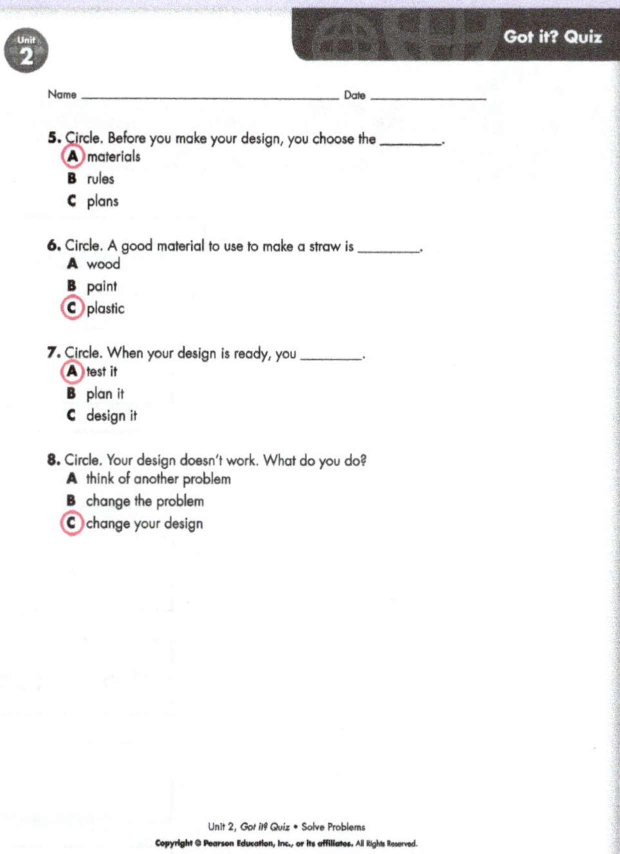

5. Circle. Before you make your design, you choose the ________.
 A materials
 B rules
 C plans

6. Circle. A good material to use to make a straw is ________.
 A wood
 B paint
 C plastic

7. Circle. When your design is ready, you ________.
 A test it
 B plan it
 C design it

8. Circle. Your design doesn't work. What do you do?
 A think of another problem
 B change the problem
 C change your design

Unit 2 • Digital Resources and Photocopiables **T27b**

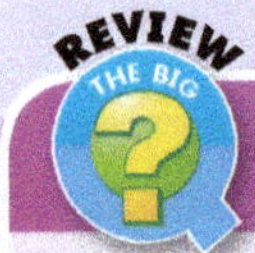

Unit 2 Study Guide

How can you solve problems?

Lesson 1
What are problems and solutions?

- A problem is something we need to fix or solve.
- A solution is an answer to a problem.

Lesson 2
How do ideas become solutions?

- We can make a plan and design our solution.
- We choose tools and materials to make our design.

Lesson 3
How can we test and share solutions?

- We test our solution. We solve many problems.
- We can share our solutions.

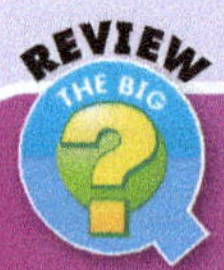

Review the Big Question

How can you solve problems?

Have students use what they have learned from the unit to answer the question in their own words.

How has your answer to the Big Question changed since the beginning of the unit? What are some things you learned that caused your answer to change?

Make a Concept Map

Draw on the board a concept map like the one shown on this page. With the students, talk through the key ideas from this unit. Invite different students to point to the ideas on the board, miming as possible.

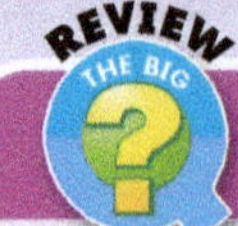

Unit 2 Concept Map

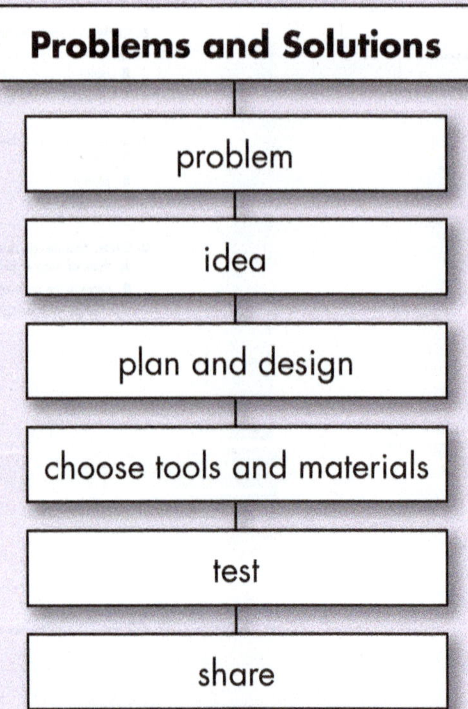

Students can make a concept map to help review the Big Question.

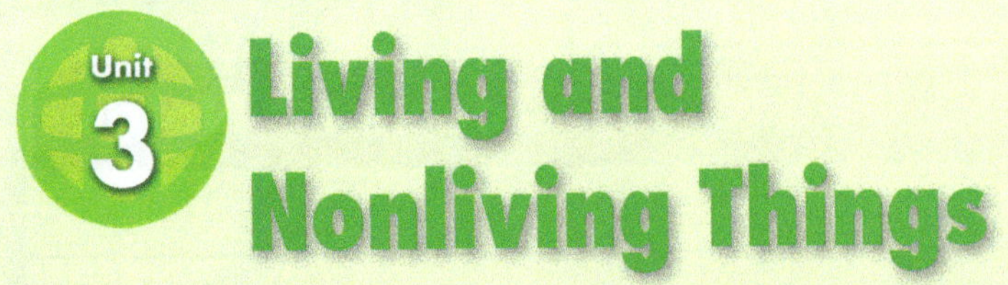

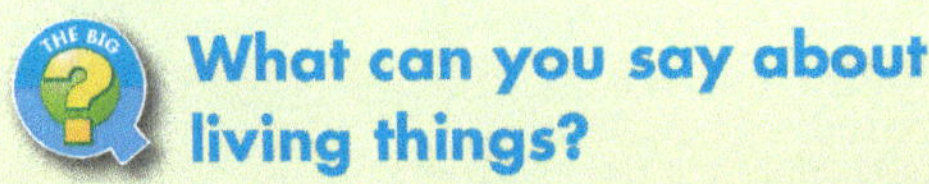

Lesson Plan

Unit Opener & Lesson 1 What are living and nonliving things?

	Activity	Pages	Time
Engage	• Unit Opener: Think! *Is the teddy bear a living thing?*	SB p. 28	5 min
	• Unit Opener: Distinguish living and nonliving things.	SB p. 28	10 min
	• Unit Opener: Identify similarities and differences.	SB p. 28	10 min
	• Think! *What is another thing you need?*	SB p. 30	5 min
Explore	• Digital Lab: *What things are living?* (ActiveTeach)	TB p. 29	30 min
Explain	• Living things grow and change	SB p. 29	30 min
	• Living things have needs	SB p. 30	30 min
	• Characteristics of nonliving things	SB p. 31	30 min
Elaborate	• Parents and Babies	TB p. 29	10 min
	• Meeting Needs	TB p. 30	20 min
	• Flash Lab: Living and Nonliving Things	SB p. 31	20 min
Evaluate	• *Lesson 1 Check* (ActiveTeach)	TB p. 39a	10 min
	• Assessment for Learning	TB p. 31	10 min
	• Review (Lesson 1)	SB p. 39	10 min
	• *Got it? Self Assessment* (ActiveTeach)	TB p. 39b	10 min
	• *Got it? Quiz* (ActiveTeach)	TB p. 39b	10 min

Lesson 2 How are animals alike and different?

	Activity	Pages	Time
Engage	• Think! *Do an animal's body parts help it get what it needs?*	TB p. 33	10 min
Explain	• Similarities and differences of some animals	SB p. 32	30 min
	• Some animal body parts and how some animals move	SB p. 33	30 min
	• Ways to group some animals	SB p. 34	30 min
Elaborate	• Feathers and Fur	TB p. 32	30 min
	• Animal Prints	TB p. 33	30 min
	• Animal Groups	TB p. 34	20 min
	• At-Home Lab: Compare Animals	SB p. 34	20 min
Evaluate	• *Lesson 2 Check* (ActiveTeach)	TB p. 39a	10 min
	• Assessment for Learning	TB p. 34	10 min
	• Review (Lesson 2)	SB p. 39	10 min
	• *Got it? Self Assessment* (ActiveTeach)	TB p. 39b	10 min
	• *Got it? Quiz* (ActiveTeach)	TB p. 39b	10 min

	Lesson 3 How are plants alike and different?		
	Activity	**Pages**	**Time**
Engage	• *Think!* *What are other different things about the cactus and the water lily?*	SB p. 36	10 min
	• *Think!* *Some plants need a lot of water. Can they be in a group?*	SB p. 37	10 min
Explain	• Ways some plants are alike and different	SB p. 35	30 min
	• Characteristics and needs of some plants	SB p. 36	30 min
	• Some ways to group plants	SB p. 37	30 min
	• *Got it? 60-Second Video* (ActiveTeach)	TB p. 25	10 min
Elaborate	• Seed Plants	TB p. 35	20 min
	• Leaf Textures	TB p. 36	30 min
	• Plants and Animals Game	TB p. 37	30 min
Evaluate	• *Lesson 3 Check* (ActiveTeach)	TB p. 39a	20 min
	• Assessment for Learning	TB p. 37	15 min
	• Review (Lesson 3)	SB p. 39	25 min
	• *Got it? Self Assessment* (ActiveTeach)	TB p. 39b	15 min
	• *Got it? Quiz* (ActiveTeach)	TB p. 39b	15 min
Lab	• *Let's Investigate! How are animals and plants different?* (ActiveTeach)	SB p. 38	30 min

Flash Cards

living/nonliving

grow

move

fin

wing

beak

leaves

flowers

seeds

Lesson 1

Key Words	ELL Support
living, grow, need, nonliving, move	**Vocabulary:** *change, air, space, water, food, on (their) own*

Lesson 2

Key Words	ELL Support
fur, body coverings, feathers, paws, fins, wings, beaks	**Vocabulary:** *move, footprints, eagle* **Body Coverings:** *hair, wool, scales* **Action Verbs:** *fly, swim, jump, run, walk*

Lesson 3

Key Words	ELL Support
stems, leaves, roots, flowers, seeds, petals, trunks	**Vocabulary:** *shape, size, color, tulips, daisies, pine tree, cactus, alike, different, oak tree, daffodils*

Unit 3
Living and Nonliving Things

Unit Objectives

Lesson 1: Students will learn about living and nonliving things and what living things need.

Lesson 2: Students will learn how animals are alike and different.

Lesson 3: Students will learn how plants are alike and different.

Vocabulary: *living, nonliving, alike, different, move, on its own, toy horse, horse, teddy bear*

Introduce the Big Question

What can you say about living things?

Build Background Say *I can walk.* and walk a few steps. Invite a student up to the front and ask *Can you walk?* Encourage the student to follow your actions. Say *I can jump.* and jump. Ask another student up to the front and ask the two students *Can you jump?* Encourage both students to jump. *We can move!* Invite all students to stand up and move in the ways you do.

Next, place a pencil on your desk. Say *Walk!* Look perplexed and ask students *Did the pencil walk? No! It can't move on its own.* Repeat for another movement and with another nonliving object students are familiar with, like a desk or backpack.

Finally, select one of the objects and make it move. Look like you have an idea and say *I can make these objects move. But, they can't move on their own because they are nonliving! We can move on our own because we are living! One thing we can say about living things is that they can move on their own!*

Engage

Think!

Is the teddy bear a living thing?

Draw students' attention to the photo of the teddy bear and ask the question. Invite students to discuss the question in pairs or small groups.

1 Circle the thing that can move on its own.

Point to the picture of the horse and elicit or tell students what it is. *Do you think it can walk? Yes! Can it jump? Yes! Can it move on its own? Yes!*

Next, point to the picture of the rocking horse. *This is a toy horse. It is made of wood. Can it walk? No! Can it move on its own? No!*

Invite students to circle the picture of the living horse.

2 How are the animals alike? Say with a partner.

Have students look at the pictures. Elicit or name the animals and invite students to say whether they are living or nonliving. Then, in pairs, students discuss how the chipmunk and the cat are alike. (Possible answers: *They have four legs. They have eyes. They have hair/fur.*)

3 How are the plants alike and different? Say as a class.

What color are the plants? Green! Are they big or small? Big! Do they have the same shape? No! What do they need to survive? Water, soil, sun.

Think! Again!

Revisit the question *Is the teddy bear a living thing?* Have the pairs or small groups share their answers with the class. Encourage students to explain their answers. (Possible answers: *I think it is nonliving. It can't move on its own.*) Accept all logical answers and provide support as needed.

What are living and nonliving things?

> **Objective:** Learn that living things grow and change.
>
> **Vocabulary:** *living, grow, change*
>
> **Digital Resources:** Flash Cards (*living/nonliving, grow*), Animal Cards (*cat and kitten, chicken and chick, cow and calf, dog and puppy, duck and duckling*), *Let's Explore!* Digital Lab

Unlock the Big Question

Write the following on the board: *I will learn about living and nonliving things. I will learn about what living things need.*

Build Background Draw students' attention to the picture of the girl and her father. *Who is big, the girl or her father? Her father!* Then have students compare the lion cub and lion. *Who is small, the baby lion or the adult lion? The baby lion! The girl and the baby lion will grow and change!*

Explore

Let's Explore! Lab   What things are living?

Objective: Learn to distinguish living and nonliving things.

Digital Resources: *Let's Explore!* Lab, *Let's Explore! Activity Card* (1 per student) (*Optional*: Do the lab in class; refer to the *Activity Card* for materials and steps.)

- Show the Digital Lab. Have students work in pairs to complete the *Activity Card*. Monitor and provide support as needed.

Explain

1   Read. Circle three living things in the picture.

Read the paragraph for students. *What does it mean to grow? Get bigger. Another thing we can say about living things is that they grow!*

Display the *living/nonliving* Flash Card and elicit which thing is living and which is nonliving. Then invite the class to say or point to three living things in the picture and then circle them individually. (Possible answers: *girl, father, horse, trees, grass.*)

2   Read again. Underline two living things with a partner.

Have students read the paragraph silently and underline two living things. Check answers as a class. (*Plants and animals.*) If students did not select the plants in exercise 1, point to the picture and invite students to repeat the exercise to ensure they

---

understand both plants and animals are living things. Point out that, even though a plant doesn't move like an animal does, it grows.

3   Look around the classroom. Point to some living things.

Give students a few seconds to look around the classroom and look for living things. (Possible answers: *classmates, teacher, plants*) If possible, ask students to look through the window and identify other living things.

4   Look at the pictures. Which is the baby? Which is the parent? Does the baby lion grow? Say with a partner.

Show the *grow* Flash Card and elicit how old the boy is in each picture. Then draw students' attention to the lions and have them compare the sizes of the two animals. Then read the questions aloud. Invite pairs to discuss. Monitor and provide support as necessary.

Then ensure students understand that *change* means *to become different.* Elicit or point out some ways the lion cub changes as it gets older. *Look, the lion has longer hair!*

Elaborate

Parents and Babies

Show the Animal Cards one by one and invite students to point to the parent and the baby. Say the names for the students to repeat. Display all the cards on the board and ask the students to discuss which baby looks most like its parent.

What are living and nonliving things?

Objective: Learn what living things need.

Vocabulary: *need, air, water, space, food*

Digital Resources: Flash Cards (*living/nonliving, grow*), *I Will Know...* Digital Activity

Materials: bottle or glass of water, sandwich or other snack (teacher use)

Build Background Mime being very hot and out of breath. *I need some water!* Drink some water. Rub your stomach and say *I'm very hungry. I need some food.* Eat the food or mime eating. *Water and food are things I need.* Explain to students that *need* means *something you have to have.*

Explain

5 With a partner, say two things living things need.

Brainstorm with the students five living things and write them on the board. Put students in pairs and ask them to choose two living things. Give them a few minutes to think about two needs each of their living things has. Have pairs share their answers with another pair. Elicit ideas for each living thing as a class.

6 Look at the picture of the dog. Draw one thing the dogs needs.

Invite students to draw something dogs need and share their drawings with the class. *Right! Dogs need food. Dogs need water!*

ELL Content Support

Like other living things, plants need food. However, plants do not eat food the way animals do. They make their own food through the process of photosynthesis: plants use light energy from the sun, carbon dioxide, and water to produce a simple sugar and oxygen. Plants can also make more complex sugars, starches, proteins, and fats. Animals that eat plants get their energy from these plant products. In turn, animals that eat plant-eating animals get their energy from the animals they eat.

Think!

What is another thing you need?

Encourage students to discuss other things they might need by saying the words or miming. They may suggest exercise or sleep, for example. Accept all logical answers.

Elaborate

Meeting Needs

Pair students and have one in each pair mime taking care of a plant or animal or themselves and the other student guess what need is being met. (*It needs water! It needs food!*) Have partners switch roles. Monitor and provide support as necessary.

I Will Know...

Have students do the *I Will Know...* Digital Activity.

What are living and nonliving things?

Objective: Learn about nonliving things.

Vocabulary: *nonliving, move, car, toy*

Digital Resources: Flash Card (*move*), *Lesson 1 Check* (print out 1 per student)

Build Background Discuss things in the classroom or visible from the classroom that are living and nonliving. Write the following sentence frames on the board.

A _____ is living.

A _____ is nonliving.

Have students draw pictures or write words on the board to complete the sentences. Provide support as necessary.

Explain

7 **Read. Look at the pictures. Circle the nonliving things.**

Read the paragraph for students. Draw their attention to the photos and elicit or identify what the objects are (*bird, building blocks, car*). Invite students to circle the nonliving things and check answers as a class. Ask questions to check comprehension. *Can the building blocks move on their own? No! Can the bird? Yes!* Show the *move* Flash Card. *Can this boy move on his own? Yes! Can you move on your own? Yes!*

8 **Look around the classroom. Point to three nonliving things with a partner.**

Have students point to three nonliving things with a partner. Invite volunteers to share their answers with the class. (Possible answers: *backpack, pencil, eraser, pen*) Provide support as needed.

9 **Does a teddy bear grow? Read and trace.**

Invite a volunteer to read the question and have students trace the word.

Follow up with some comprehension questions. *Does a teddy bear need to eat? No! Does a toy need to drink water? No! Does a car grow bigger? No!*

---

7 Read. Look at the pictures. Circle the nonliving things.

Nonliving Things

Nonliving things do not grow. Nonliving things do not change. Nonliving things cannot **move** on their own. A car is a nonliving thing. A toy is a nonliving thing.

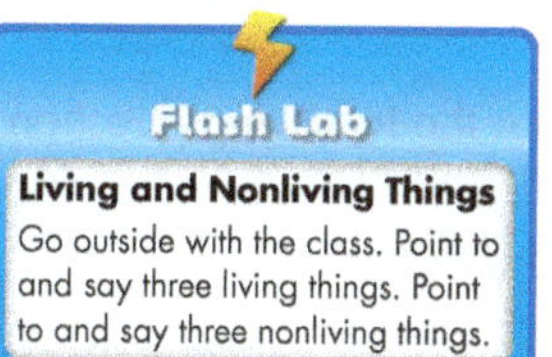

8 Look around the classroom. Point to three nonliving things with a partner.

9 Does a teddy bear grow? Read and trace.

No. A teddy bear is nonliving.

Flash Lab

Living and Nonliving Things

Go outside with the class. Point to and say three living things. Point to and say three nonliving things.

Lesson 1 Check | Unit 3 | 31

Elaborate

⚡ Flash Lab

Living and Nonliving Things

Take the class outside and have students identify three living and three nonliving things. Encourage students to say how they know. (Possible answers: *It can move on its own. It doesn't grow.*) Monitor and provide vocabulary support as needed.

Evaluate

Lesson 1 Check Assessment for Learning

Review the Key Words for Lesson 1 (see Student's Book page 29). Distribute the *Lesson 1 Check* and guide students as they complete it. Check answers as a class. Then ask students to grade their progress on the topic of living and nonliving things from 1 to 3: 3 = *I understand what living and nonliving things are;* 2 = *I need to study more;* 1 = *I need help!* Encourage students giving themselves a 1 or a 2 to say what they found difficult and what they need to study more.

Lesson 2

How are animals alike and different?

Objective: Learn how to compare some animals.

Vocabulary: *fur, body coverings, feathers, wings, beak, sounds, colors, shapes*

Digital Resources: Flash Cards (*wings, beak*)

Materials: art supplies, fake fur, feathers

Unlock the Big Question

Write the following question on the board: *I will learn how some animals are alike and different.*

Build Background Activate prior knowledge by having students think about a pet or any animals they have seen where they live. Have them draw a quick sketch of the animal. Invite volunteers to share their pictures. As a class, discuss some similarities and differences between two pictures at a time. (Possible answers: *This animal has two legs. This one has four legs.*) *We're going to learn some more ways animals can be the same and different in this lesson!*

Explain

1 **Read. Underline one way cats and dogs are alike.**

Read the paragraph aloud for students. Invite them to identify one way cats and dogs are alike. *They both have fur.* To aid comprehension, have students look at the pictures of the dog on page 30 and cat on page 28 to see the fur. Ask if they can point to other animals in this unit that have fur. (Possible answers: *chipmunk, baby lion, lion*)

Then talk about the differences. Elicit or say the sounds cats and dogs make to reinforce comprehension. (*Meow! Woof!*)

ELL Vocabulary Support

You may wish to take the opportunity to teach some other body coverings like *hair, scales,* and *wool. Sheep have wool! Fish have scales!* You also may wish to take some time to teach some action verbs like *fly, jump, swim, run, walk,* and so on.

Lesson 2 · How are animals alike and different?

Key Words
- fur
- body coverings
- feathers
- paws
- fins
- wings
- beaks

1 **Read. Underline one way cats and dogs are alike.**

Compare Animals

Animals can be alike. Cats and dogs both have fur. Animals can be different. Cats and dogs make different sounds. Animals can have different colors and shapes. They can have different **body coverings.** Birds don't have fur. They have **feathers.**

2 **Look at the pictures. Say one way the animals are alike with a partner.**

3 **Look at the pictures again. With a partner, say two ways the animals are different.**

32   Unit 3

2 **Look at the pictures. Say one way the animals are alike with a partner.**

Reread the last few sentences of the paragraph. Explain that *body covering* means what is on the body of an animal. Point and say *Look, I have hair on my body!* Explain that some animals have hair or fur, some animals have feathers, and some animals have other body coverings. For example, snakes and fish have scales.

Display the *beak* Flash Card and say the word for the students to repeat. Invite students to say two ways the birds are alike. Accept all logical answers. Ensure students notice that the birds have feathers covering their bodies.

3 **Look at the pictures again. With a partner, say two ways the animals are different.**

Invite pairs to discuss the differences among the birds. (Possible answers: *They are different colors. They are different sizes. Their beaks are different shapes.*)

Elaborate

Feathers and Fur

Have students complete their sketches from the beginning of the class and color them. Invite students to add feathers or bits of fur as appropriate. Alternatively, students can draw to indicate other body coverings, like scales. Have students compare their animals with a partner and describe some ways their animals are the same and different. Ensure they describe the body coverings their animal has. Provide support as needed.

How are animals alike and different?

Objective: Learn about some animal body parts.

Vocabulary: *paws, fins, wings, beaks, kitten, ant, bee, elephant, frog, swim, fly, jump*

Digital Resources: Flash Cards (*fins, wings, beak*), Animal Card (*eagle*), *I Will Know…* Digital Activity

Materials: pictures of paw prints, bird footprints, elephant footprints, art paper

Build Background Display the Flash Cards and elicit the two kinds of animal. (Answer: *fish, bird*) Pre-teach *fins.* Invite students to find differences and similarities among the three animals. Help students review the words from the previous lesson (*feathers, scales, wings, beak*).

Explain

4 Read. With a partner, say a body part fish have.

Read the paragraph with students. Ask the question and have students answer. *Right! Fish have fins! Do people have fins? No!*

Point to the pictures of the birds on page 32. Point to the label showing the birds' wings. Then display the *eagle* Animal Card and show what the eagle's wings look like.

5 Look at the pictures. What animal has two legs? Circle.

Point to and identify the animals in the pictures. Have students circle which animal has two legs and have pairs check answers.

Follow up with some comprehension questions. *Does the elephant have feathers? No! Does the bird have feathers? Yes! Does the cat have paws? Yes! Is your mouth like a bird's mouth? No, birds have beaks!*

6 Read. Underline three ways that some animals can move.

Living things can move on their own. But, some animals move in different ways from other animals! Read the paragraph. Have students mime with you the ways the animals move. Have students underline the ways the animals pictured move. Show the *eagle* Animal Card again and say *This is what the eagle's wings look like when it flies.*

Then ask what body parts each animal they have learned about in this lesson has. (Possible answers: *The fish have fins. The bee has wings.*)

4 Read. With a partner, say a body part fish have.

Animal Body Parts

Some animals have the same body parts. Dogs and cats have **paws**. Some animals have different body parts. Fish and birds don't have paws. Fish have **fins**. Birds have **wings**. They also have **beaks**.

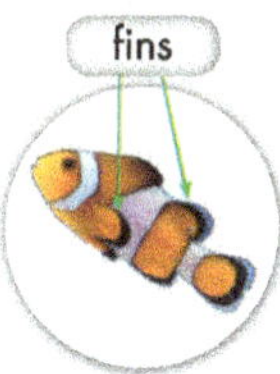

5 Look at the pictures. What animal has two legs? Circle.

6 Read. Underline three ways that some animals can move.

How Animals Move

Animals move in different ways. Some animals <u>swim.</u> Fish swim. Some animals <u>fly.</u> Bees fly. Some animals <u>jump.</u> Frogs jump.

I Will Know… Unit 3 **33**

Elaborate

Animal Prints

Distribute paper and pencils to all students. Pair students and help them draw an outline of one another's foot at one end of their paper. Then show the pictures of the animal footprints. *This is what an elephant's footprint looks like.* Have students draw what the other animals' footprints look like next to their own.

Think!

Do an animal's body parts help it get what it needs?

Read the question aloud for students and lead a class discussion. (Possible answers: *I think a bird's beak helps it get food. I think its eyes help it find food.*) Accept all logical answers. Provide vocabulary support as needed.

I Will Know…

Have students do the *I Will Know…* Digital Activity.

How are animals alike and different?

> **Objective:** Learn how to group animals.
>
> **Vocabulary:** *fur, group*
>
> **Digital Resources:** 1 set of Animal Cards per small group (*panda, monkey, dog and puppy, cat and kitten, eagle, firefly, pelican, kangaroo, frog, rabbit*), *Lesson 2 Check* (print out 1 per student)

Build Background Activate prior knowledge by discussing with students how scientists put things in groups. *They put things that are alike into a group.*

Explain

7 Read. With the class, say two animals that can fly.

Read the paragraph with students. Have students name two animals that can fly. (Possible answers: *birds, bees*)

8 Look at the pictures. Make a group of animals. Circle the animals that are alike.

Direct students' attention to the pictures and have them identify the animals. Then have them make a group by circling the animals that are alike. Check answers as a class. Guide students to answer that the dog and the cat both have fur.

ELL Vocabulary Support

Help students remember new vocabulary by using simple diagrams. Draw spidergrams on the board like the ones below:

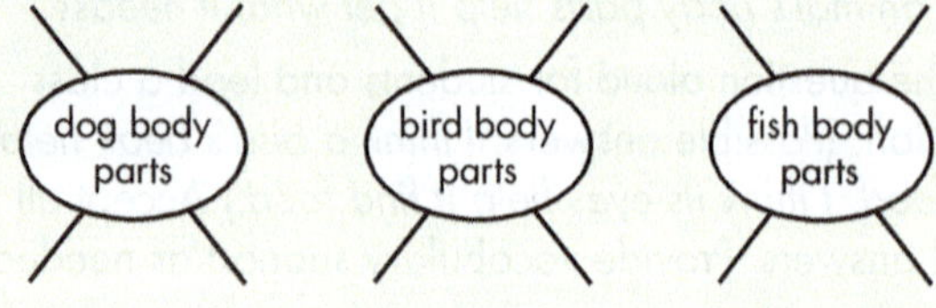

Elicit words from the students and add them to the spidergrams. You may wish to copy the spidergrams on cards and use them to review vocabulary.

9 Draw one animal that can swim. Share your drawing with the class.

Brainstorm with students the animals they know that can swim. Students may mime, say the sounds the animals make, or say the names of the animals. (Possible answers: *fish, dogs, ducks*) Accept all logical answers and provide vocabulary support as needed. Invite students to draw an animal that can swim and to share their drawings with the class.

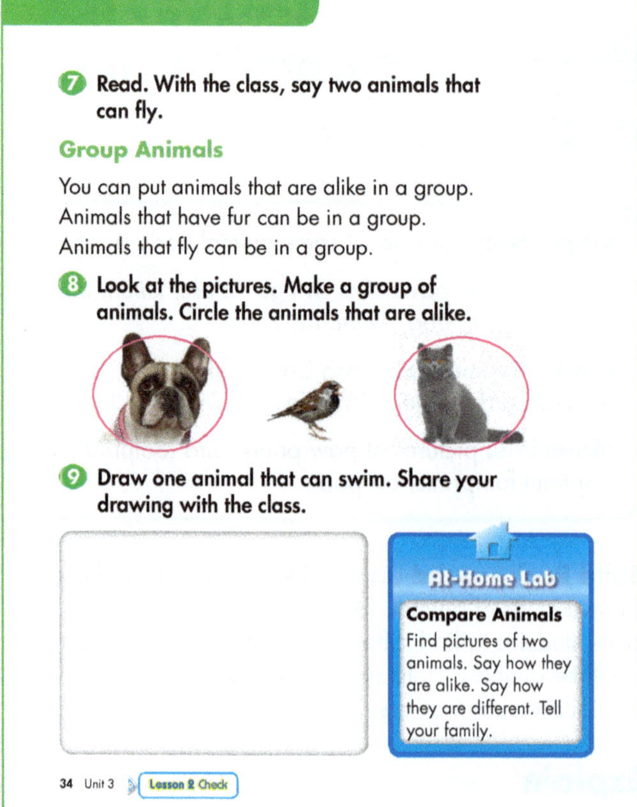

7 Read. With the class, say two animals that can fly.

Group Animals

You can put animals that are alike in a group. Animals that have fur can be in a group. Animals that fly can be in a group.

8 Look at the pictures. Make a group of animals. Circle the animals that are alike.

9 Draw one animal that can swim. Share your drawing with the class.

At-Home Lab

Compare Animals
Find pictures of two animals. Say how they are alike. Say how they are different. Tell your family.

34 Unit 3   Lesson 2 Check

Elaborate

Animal Groups

Distribute Animal Cards, mixed up, to groups. Invite groups to make groups of animals. Help groups to make at least two groups (animals that have fur, can fly, can jump, have feathers, have paws, etc.). Accept all logical answers.

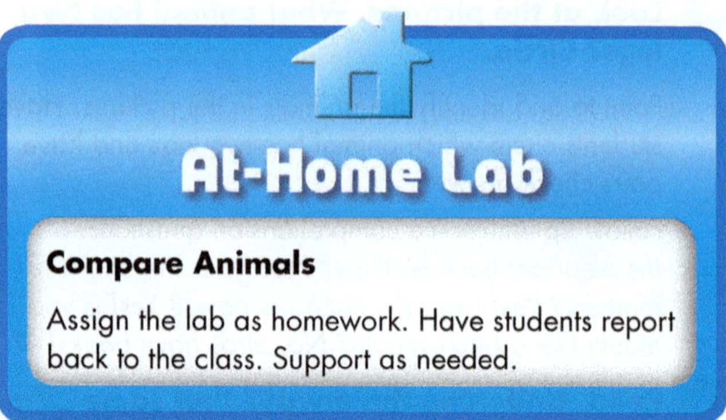

At-Home Lab

Compare Animals

Assign the lab as homework. Have students report back to the class. Support as needed.

Evaluate

Lesson 2 Check Assessment for Learning

Review the Key Words for Lesson 2 (see Student's Book page 32). Distribute the *Lesson 2 Check* and guide students as they complete it. Check answers as a class. Then ask students to grade their progress on the topic of how animals are alike and different from 1 to 3: 3 = *I understand some ways animals are alike and different;* 2 = *I need to study more;* 1 = *I need help!* Encourage students giving themselves a 1 or a 2 to say what they found difficult and what they need to study more.

How are plants alike and different?

Objective: Learn how plants are alike and different.

Vocabulary: *plants, stems, leaves, roots, flowers, seeds, soil, tulips, daisies, pine tree, cactus*

Digital Resources: Flash Cards (*leaves, flowers, seeds*)

Materials: plant (ideally with flowers), a picture of a sunflower, a handful of sunflower seeds in the shell

Unlock the Big Question

Write the following text on the board: *I will learn how plants are alike and different.* Explain to students that, like animals, plants can be alike and different and that they will learn more about this in this lesson.

Build Background Display the plant on a table and invite students to gather around. Point to different parts of the plant and invite students to describe them. Provide support as needed. (*This part is green. This part is (pink). This part is long and tall. It's smooth.*)

Explain

1 **Read. What are three ways plants can be alike? Say as a class.**

Read the paragraph aloud for students. Display the Flash Cards and use them to point out the highlighted words. Have students repeat after you.

Direct students' attention to the labeled photo and invite volunteers to read the different parts of the plant. Have the class say three ways plants can be alike.

Display the plant again and follow up with some comprehension questions. *What parts does this plant have?* If necessary, explain that the plant's roots are in the soil.

2 **Look at the plants. With a partner, say how they are alike.**

Invite students to look at the pictures of the tulips and daisies. Have pairs discuss how they are alike. Have pairs share their answers with another pair. (Possible answers: *They have flowers. They have leaves. They have stems.*) Monitor and provide support as necessary. Then invite students to say how the plants are different. (*They are different colors. Their petals are different shapes.*)

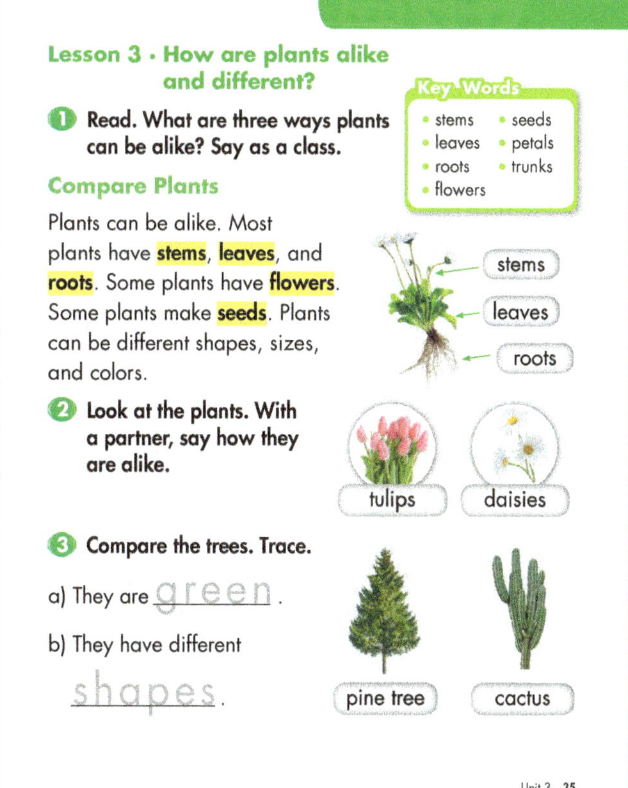

3 **Compare the trees. Trace.**

Have students look at the pictures of the plants and compare them by filling in the sentences. Elicit any other differences. (Possible answers: *A cactus doesn't need a lot of water to grow. A pine tree smells nice. A cactus grows in the desert. A cactus has flowers. You can get wood from a pine tree.*)

Elaborate

Seed Plants

Display the sunflower seeds you brought on a table and gather students around. Explain that some plants have seeds and that these seeds come from sunflowers. Show the picture of the sunflower and point out where the seeds grow on a sunflower. Then show students how the seed is inside the shell. Invite students to say why they think plants have seeds and why they think the sunflower seeds might sit inside a shell. Accept all logical answers and provide support as needed. (Possible answers: *Plants have seeds so that they can make new plants of the same kind. Each seed can make a new plant. The sunflower seeds sit inside a shell to keep them safe. The shell makes it harder for animals to eat the seed.*)

How are plants alike and different?

> **Objective:** Learn about more plant parts and the needs of some plants.
>
> **Vocabulary:** *tulips, daisies, stems, leaves, petals, wide, thin, water lilies, cactuses*
>
> **Digital Resources:** *I Will Know…* Digital Activity
>
> **Materials:** flower petals of different widths, leaves of different sizes, shapes, and textures

Build Background Invite students to draw a flower. Invite volunteers to show their pictures to the class. Provide support as needed. Have students note similarities and differences. (*This flower is red. This one is pink. The leaves are big.*)

Explain

4 **Look at the pictures. How are the flowers different? Say as a class.**

Read the paragraph aloud to students. Invite a volunteer to point to the thin and wide petals.

Display the petals you brought to class. Invite students to describe them. (Possible answers: *They are yellow and wide. They are purple and thin.*) Then have volunteers describe the width of the petals on the flowers they drew.

5 **Read. What plant needs a lot of water? Mark (✔).**

Read the paragraph with students and have them mark the plant that needs a lot of water. Have students recall what living things need. Explain that all plants need water, but some need more than others. Invite students to think about whether the plants they drew need a lot of water.

ELL Content Support

For most plants, their parts include leaves, stems, and roots. These parts can vary greatly from one kind of plant to another. For example, water lilies have roots that are in the mud at the bottom of the body of water where they grow. The sharp spines of cactuses are modified leaves.

---

4 **Look at the pictures. How are the flowers different? Say as a class.**

Petals

Tulips and daisies are alike. They have stems and leaves. They make flowers. But they are different, too. Tulip flowers have wide **petals**. Daisy flowers have thin petals.

5 **Read. What plant needs a lot of water? Mark (✔).**

Compare Needs

All plants need water. Some plants need a lot of water. Water lilies need a lot of water. Some plants don't need a lot of water. Cactuses don't need a lot of water.

36 Unit 3   I Will Know…

Think!

What are other different things about the cactus and the water lily?

Ask the question and have students say or point to differences they can see in the photos. (Possible answers: *The water lily has flowers and petals. The cactus has fruit and pointy spines. They are different colors. The water lily lives in water. The cactus lives where it's hot.*)

Elaborate

Leaf Textures

Distribute leaves to students. Show them how to place a piece of paper on top of their leaf and rub the pencil back and forth to trace the shape and transfer the texture of the leaves. Have students notice the differences in shapes, sizes, and, where possible, textures. Provide vocabulary support as necessary.

> **I Will Know…**
> Have students do the *I Will Know…* Digital Activity.

How are plants alike and different?

> **Objective:** Learn about grouping plants.
>
> **Vocabulary:** *daffodils, pine tree, oak tree*
>
> **Digital Resources:** *Lesson 3 Check* (print out 1 per student), *Got it? 60-Second Video*
>
> **Materials:** several pairs of cards (one that pictures a plant and one an animal per pair of cards)

Build Background Activate prior knowledge by having students recall the common parts of plants (*stem, roots, leaves*). Have them look at the pictures of the plants on the page and identify the stems where possible. Explain that a trunk is a stem that is hard and woody. Invite students to say which plants on the page are trees.

Explain

6 Read. Can trees be in a group? Talk as a class.

Read the paragraph along with students and lead a discussion about whether trees can be in a group. Guide students to conclude that they can.

ELL Content Support

The trunk of a tree sends water and nutrients from the roots to the branches and the leaves. The outside layer, called bark, is hard so that it protects the trunk from extreme temperatures, bad weather, insects, and fungi.

7 Look at the pictures. Circle the plants that can be in a group. Say why with a partner.

Direct students' attention to the photos and have students circle the plants that can be in a group. Check answers as a class.

Follow up by having students make two groups of plants from the pictures on the page, trees and flowers.

Elaborate

Plants and Animals Game

Divide the class into two teams, one on each side of the classroom. For each team, write *Plants* and *Animals* on the board. Hand out an equal number of card pairs, face down. Have the first person in each team turn over a pair of cards at the same time and put each card under the appropriate heading on the board. The first team to correctly categorize their card pairs wins.

Think!

Some plants need a lot of water. Can they be in a group?

Read the question aloud for students. Allow students time to discuss freely. Guide students to conclude that plants that need a lot of water are alike in that way, so they can be in a group.

Evaluate

Lesson 3 Check   Assessment for Learning

Review the Key Words for Lesson 3 (see Student's Book page 35). Distribute the *Lesson 3 Check* and guide students as they complete it. Check answers as a class. Then ask students to grade their progress on the topic of how plants are alike and different from 1 to 3: 3 = *I understand some ways plants are alike and different;* 2 = *I need to study more;* 1 = *I need help!* Encourage students giving themselves a 1 or a 2 to say what they found difficult and what they need to study more.

Got it?   60-Second Video

Play the *Got it? 60-Second Video* to review the unit material.

Let's Investigate!

In this unit, students learn about some ways animals are alike and different and some ways plants are alike and different. In this lab, they will compare some animals and some plants.

Let's Investigate! Lab **How are animals and plants different?**

Objective: Students will learn how to compare some animals and some plants.

Materials: paper, pencil, picture of a dog, a bird, a tree (without flowers), and a daisy plant.

Digital Resources: *Let's Investigate!* Digital Lab, *Let's Investigate! Activity Card* (1 per group)

- Display the picture of the dog and the bird. Have students discuss what they look like. Ask questions to help students think about their differences. *How many legs do they have? Does the dog have a beak? Does the dog have wings?*

- Next, distribute the *Activity Cards* and invite students to draw the two animals, emphasizing their differences. Monitor and provide support as needed.

- Repeat the procedure for the plants.

- Invite a few volunteers to show their pictures and have the class discuss the differences between the animals and between the plants.

Teacher Time-Saving Option: Show the *Let's Investigate!* Digital Lab as an alternative to the hands-on lab activity.

Unlock the Big Question

Have students refer to the Big Question on the Unit Opener page. In pairs have them recall what they have learned about living and nonliving things and some differences between animals and between plants. Have students complete questions 5 and 6 on the *Activity Card*.

Let's Investigate!

How are animals and plants different?

1. Look at the animals.
2. Compare them.
3. Look at the plants.
4. Compare them.

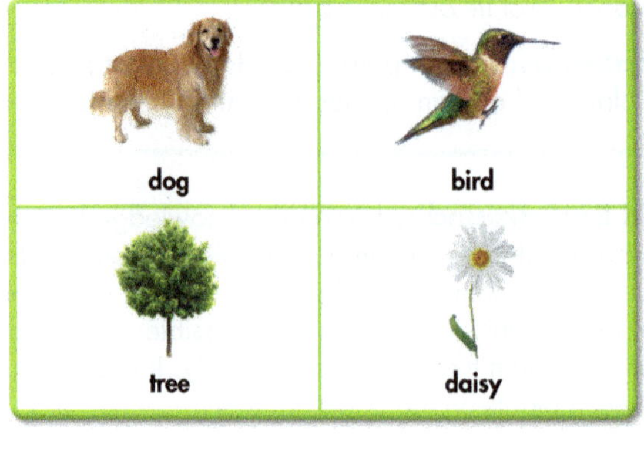

Class Project: Living and Nonliving Things

Materials: art supplies, magazines with pictures of plants, animals, and inanimate objects, safety scissors

Divide the class into small groups. Distribute materials. Instruct groups to look for pictures of living and nonliving things and to put them into groups. Allow adequate time for students to collect and cut out the pictures. Then have students group the living things into plants and animals. Write *Living Things, Nonliving Things, Plants,* and *Animals* on the board. Have students copy *Living Things* at the top of one sheet of paper and *Nonliving* at the top of the other. Have them copy *Plants* on the left and *Animals* on the right underneath *Living*. Invite students to glue their pictures on the appropriate parts of the pages. Invite groups to present their collages to the class and name as many of the items as they can. Provide vocabulary support as needed. Display students' work around the classroom.

Unit 3 Review

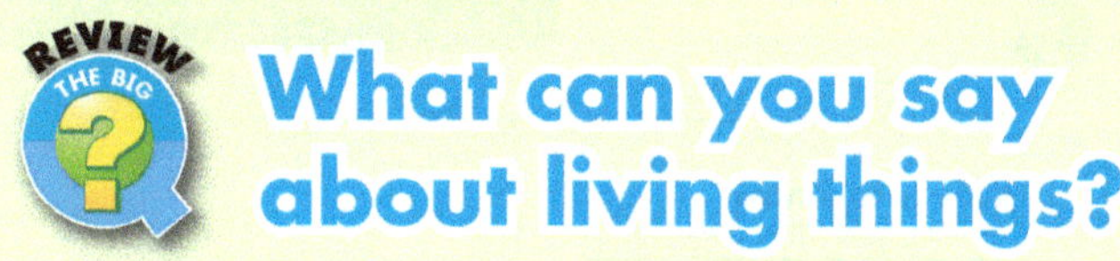

What can you say about living things?

Digital Resources: Print out 1 of each per student: *Got it? Self Assessment, Got it? Quiz*

Evaluate

Strategies for Targeted Review

The following are strategies for providing targeted review for students if they encounter challenges with the content.

Lesson 1 What are living and nonliving things?

Question 1

If... students are having difficulty matching the girl to the things she needs, then... mime being really thirsty and hungry. Then ask students if living things need teddy bears.

Lesson 2 How are animals alike and different?

Question 2

If... students are having difficulty circling how the animals are alike, then... review animal body coverings and body parts. *Do dogs fly? No!*

Lesson 3 How are plants alike and different?

Question 3

If... students are having difficulty distinguishing true and false answers, then... invite students to review the lesson on plants' similarities and differences.

ELL Language Support

Before students start working on the Review activities, have them read each question aloud along with you.

Got it? Self Assessment

Immediately after students have completed the Review activities, distribute a *Got it? Self Assessment* to each student. Have students complete the *Stop! Wait!* and *Go!* statements for each lesson, allowing them to look back through the lesson material if necessary.

Got it? Quiz

Distribute a Unit 3 *Got it? Quiz* to each student. Quizzes may be used for assessing students' understanding of unit concepts as well as for grading purposes.

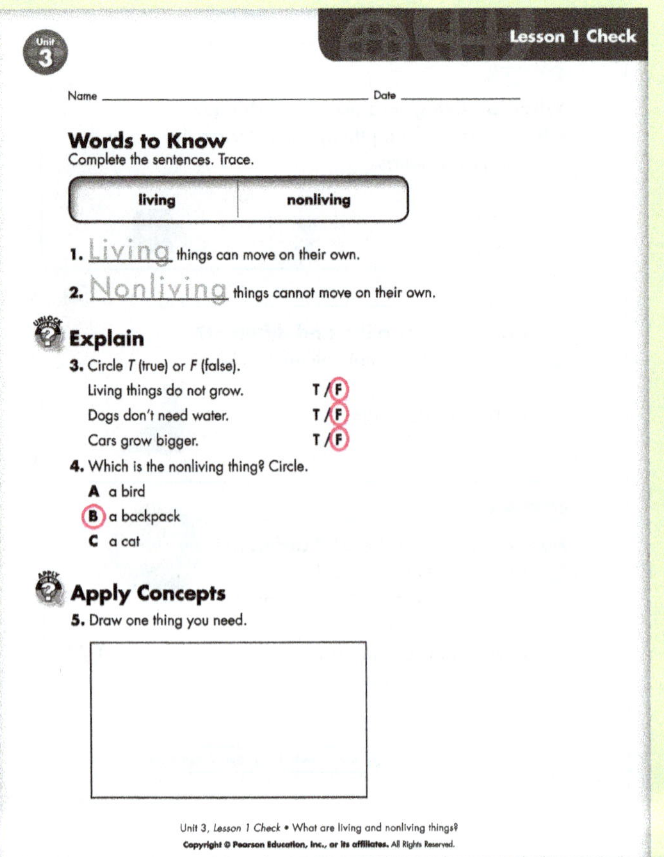

Unit 3
Lesson 1 Check
Name ____________ Date ____________
Words to Know
Complete the sentences. Trace.
living | nonliving
1. Living things can move on their own.
2. Nonliving things cannot move on their own.
Explain
3. Circle T (true) or F (false).
Living things do not grow. T / F
Dogs don't need water. T / F
Cars grow bigger. T / F
4. Which is the nonliving thing? Circle.
A a bird
B a backpack
C a cat
Apply Concepts
5. Draw one thing you need.
Unit 3, Lesson 1 Check • What are living and nonliving things?
Copyright © Pearson Education, Inc., or its affiliates. All Rights Reserved.

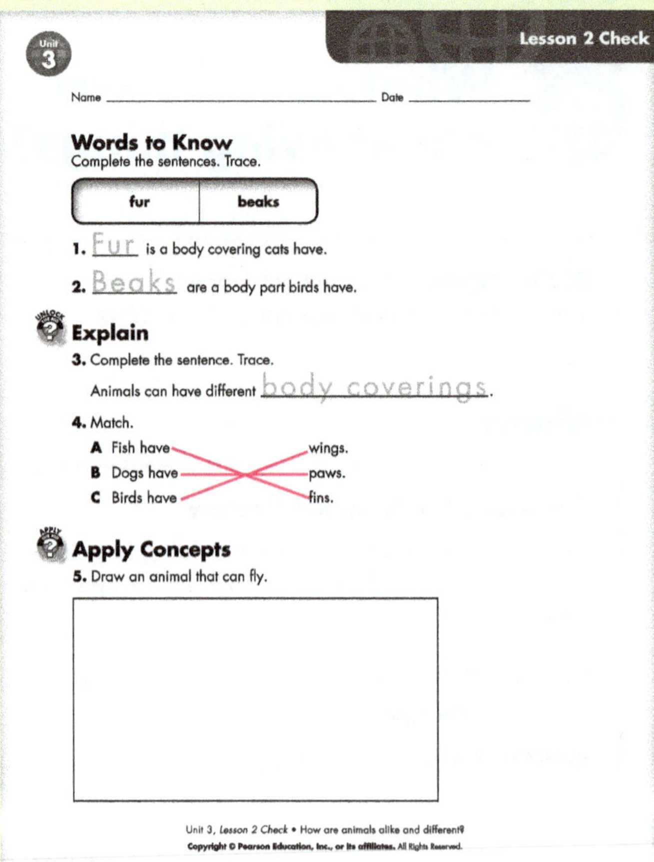

Unit 3
Lesson 2 Check
Name ____________ Date ____________
Words to Know
Complete the sentences. Trace.
fur | beaks
1. Fur is a body covering cats have.
2. Beaks are a body part birds have.
Explain
3. Complete the sentence. Trace.
Animals can have different body coverings.
4. Match.
A Fish have — wings.
B Dogs have — paws.
C Birds have — fins.
Apply Concepts
5. Draw an animal that can fly.
Unit 3, Lesson 2 Check • How are animals alike and different?
Copyright © Pearson Education, Inc., or its affiliates. All Rights Reserved.

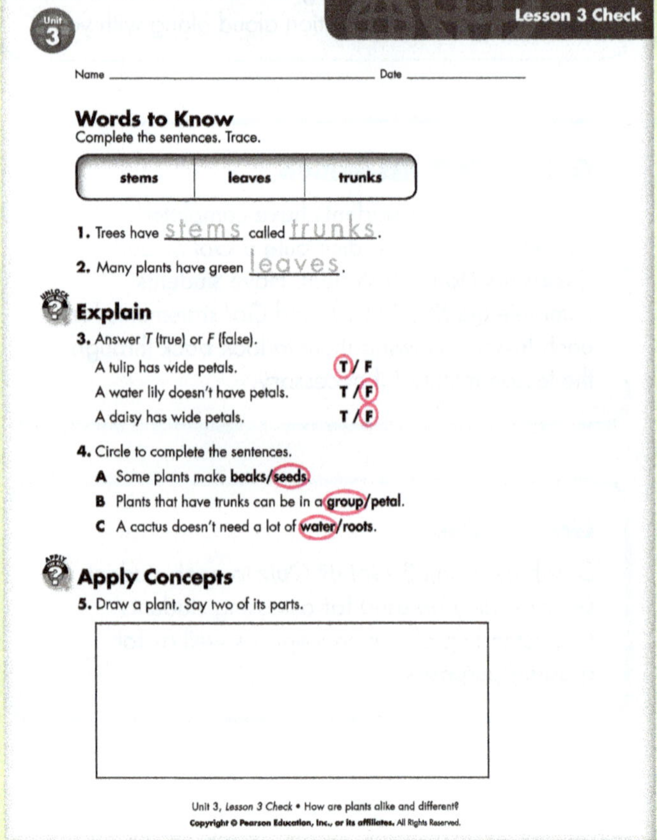

Unit 3
Lesson 3 Check
Name ____________ Date ____________
Words to Know
Complete the sentences. Trace.
stems | leaves | trunks
1. Trees have stems called trunks.
2. Many plants have green leaves.
Explain
3. Answer T (true) or F (false).
A tulip has wide petals. T / F
A water lily doesn't have petals. T / F
A daisy has wide petals. T / F
4. Circle to complete the sentences.
A Some plants make beaks/seeds.
B Plants that have trunks can be in a group/petal.
C A cactus doesn't need a lot of water/roots.
Apply Concepts
5. Draw a plant. Say two of its parts.
Unit 3, Lesson 3 Check • How are plants alike and different?
Copyright © Pearson Education, Inc., or its affiliates. All Rights Reserved.

Unit 3
Lesson 1 Let's Explore! Activity Card
Name ____________ Date ____________
Materials
• common objects, e.g., safety scissors, wooden block, ball, snap cube, box of tissues, potted plant
What things are living?
1. Look.
2. Tell.
3. Draw.
Living | Nonliving
Unit 3, Lesson 1 Let's Explore! Lab • What are living and nonliving things?
Copyright © Pearson Education, Inc., or its affiliates. All Rights Reserved.

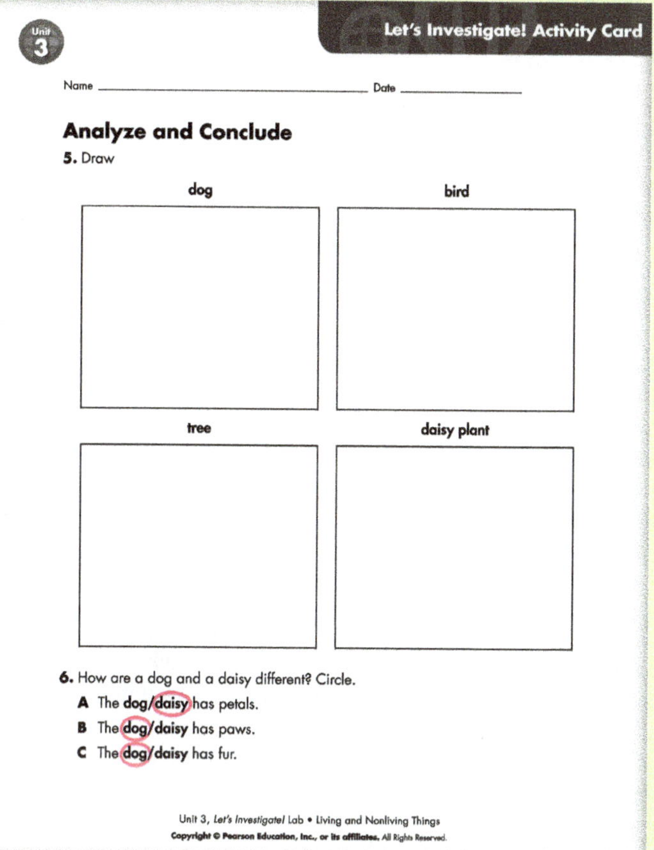

Unit 3
Let's Investigate! Activity Card
Name ___________ Date ___________
Analyze and Conclude
5. Draw
dog
bird
tree
daisy plant
6. How are a dog and a daisy different? Circle.
A The dog/daisy has petals.
B The dog/daisy has paws.
C The dog/daisy has fur.
Unit 3, Let's Investigate! Lab • Living and Nonliving Things
Copyright © Pearson Education, Inc., or its affiliates. All Rights Reserved.

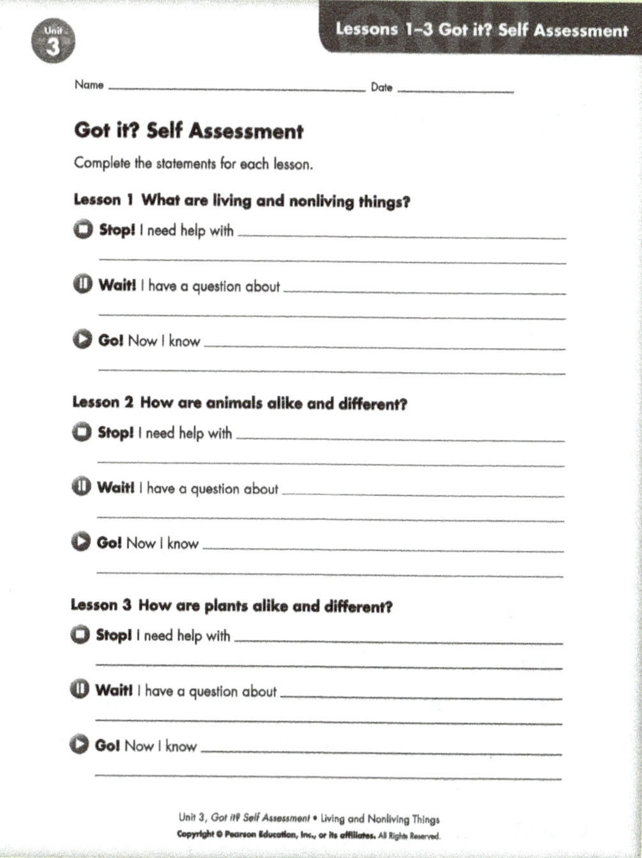

Unit 3
Lessons 1–3 Got it? Self Assessment
Name ___________ Date ___________
Got it? Self Assessment
Complete the statements for each lesson.
Lesson 1 What are living and nonliving things?
Stop! I need help with ___________
Wait! I have a question about ___________
Go! Now I know ___________
Lesson 2 How are animals alike and different?
Stop! I need help with ___________
Wait! I have a question about ___________
Go! Now I know ___________
Lesson 3 How are plants alike and different?
Stop! I need help with ___________
Wait! I have a question about ___________
Go! Now I know ___________
Unit 3, Got it? Self Assessment • Living and Nonliving Things
Copyright © Pearson Education, Inc., or its affiliates. All Rights Reserved.

Unit 3
Got it? Quiz
Name ___________ Date ___________
Got it? Quiz
1. Circle the nonliving thing.
A backpack
B daisy plant
C bird
2. Circle the living thing.
A backpack
B daisy plant
C teddy bear
3. Circle what all living things do.
A grow
B fly
C swim
4. What do living things need? Circle.
A a car
B water
C candy
Unit 3, Got it? Quiz • Living and Nonliving Things
Copyright © Pearson Education, Inc., or its affiliates. All Rights Reserved.

Unit 3
Got it? Quiz
Name ___________ Date ___________
5. Which animal has fur? Circle.
A fish
B frog
C monkey
6. Frogs can move on their own. They can ________. Circle.
A fly
B jump
C whistle
7. Plants ________. Circle.
A grow
B walk
C fly
8. These can be in a group. Circle.
A daisies and tulips
B beaks and oak trees
C frogs and trunks
Unit 3, Got it? Quiz • Living and Nonliving Things
Copyright © Pearson Education, Inc., or its affiliates. All Rights Reserved.

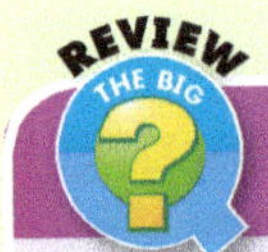

Unit 3 Study Guide

What can you say about living things?

Lesson 1
What are living and nonliving things?

- Living things grow, change, and move on their own.
- Living things have needs.

Lesson 2
How are animals alike and different?

- Animals are living things.
- Some animals have different body coverings.
- Some animals have different body parts.

Lesson 3
How are plants alike and different?

- Plants are living things.
- Most plants have stems, leaves, and roots.
- Some plants have petals. Some have trunks.

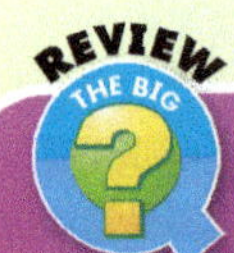

Review the Big Question

What can you say about living things?

Have students use what they have learned from the unit to answer the question in their own words.

How has your answer to the Big Question changed since the beginning of the unit? What are some things you learned that caused your answer to change?

Make a Concept Map

Draw on the board a concept map like the one shown on this page. With the students, talk through the key ideas from this unit. Invite different students to point to the ideas on the board, miming as possible.

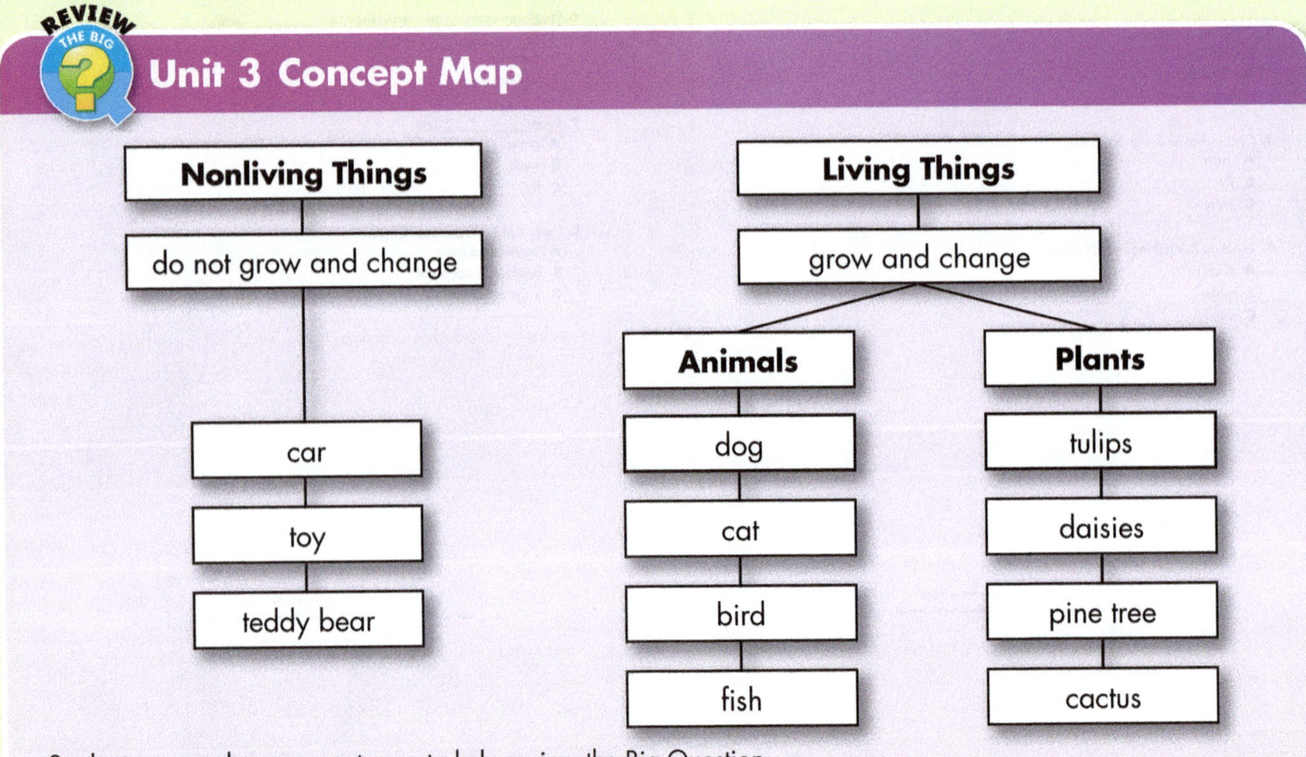

Students can make a concept map to help review the Big Question.

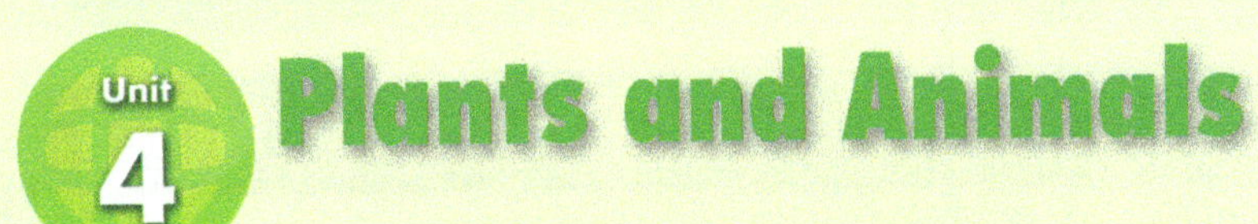
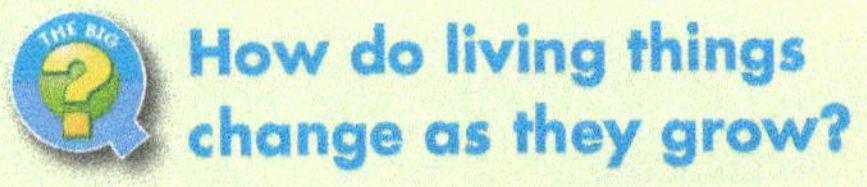

Lesson Plan

Unit Opener & Lesson 1 Do all young animals look like their parents?			
	Activity	**Pages**	**Time**
Engage	• Unit Opener: Think! *How are you like your parents? How are you different from your parents?*	SB p. 40	10 min
	• Unit Opener: Distinguish animals that look like their parents and animals that don't look like their parents.	SB p. 40	10 min
	• Unit Opener: Identify young and adult plants.	SB p. 40	10 min
	• Think! *Why can't baby birds fly?*	TB p. 41	10 min
Explain	• Some animals look like their parents and others don't	SB p. 41	30 min
	• How people grow and change	SB p. 42	30 min
	• Some animals come from eggs	SB p. 43	30 min
Elaborate	• Baby Animals	TB p. 41	20 min
	• At-Home Lab: Grow and Change	SB p. 42	20 min
	• Most Important Change	TB p. 42	20 min
	• What animal egg is it?	TB p. 43	10 min
Evaluate	• *Lesson 1 Check* (Active Teach)	TB p. 51a	10 min
	• Assessment for Learning	TB p. 43	10 min
	• Review (Lesson 1)	SB p. 51	10 min
	• *Got it? Self Assessment* (Active Teach)	TB p. 51b	10 min
	• *Got it? Quiz* (Active Teach)	TB p. 51b	10 min

Lesson 2 How do some animals grow and change?			
	Activity	**Pages**	**Time**
Engage	• Think! *Do frogs always complete their life cycles?*	TB p. 44	10 min
Explain	• Life cycle of a frog	SB p. 44	30 min
	• Life cycle of a butterfly	SB p. 45	30 min
	• How different animals grow and change	SB p. 46	30 min
Elaborate	• Worm and Fish Life Cycles	TB p. 44	20 min
	• Make a Frog's Life Cycle Poster	TB p. 45	30 min
	• Animal Groups	TB p. 46	20 min
	• At-Home Lab: Grow and Change	SB p. 46	20 min
Evaluate	• *Lesson 2 Check* (ActiveTeach)	TB p. 51a	10 min
	• Assessment for Learning	TB p. 46	10 min
	• Review (Lesson 2)	SB p. 51	10 min
	• *Got it? Self Assessment* (ActiveTeach)	TB p. 51b	10 min
	• *Got it? Quiz* (ActiveTeach)	TB p. 51b	10 min

	Lesson 3 How do some plants grow and change?		
	Activity	**Pages**	**Time**
Engage	• Think! *Do all plants have flowers?* • Think! *Why are there different types of fruit seeds?*	SB p. 47 TB p. 48	10 min 10 min
Explore	• Digital Lab: *What things are living?* (ActiveTeach)	TB p. 47	30 min
Explain	• Life cycle of some plants • Life cycle of an apple tree • Life cycle of a bean plant • *Got it? 60-Second Video*	SB p. 47 SB p. 48 SB p. 49 TB p. 49	30 min 30 min 30 min 10 min
Elaborate	• Jack and the Giant Beanstalk • At-Home Lab: *Does it have seeds?*	TB p. 49 TB p. 49	10 min 20 min
Evaluate	• *Lesson 3 Check* (ActiveTeach) • Assessment for Learning • Review (Lesson 3) • *Got it? Self Assessment* (ActiveTeach) • *Got it? Quiz* (ActiveTeach)	TB p. 51a TB p. 49 SB p. 51 TB p. 51b TB p. 51b	10 min 10 min 10 min 10 min 10 min
Lab	• *Let's Investigate! How does a butterfly change?* (ActiveTeach)	SB p. 50	30 min

Flash Cards

look like

butterfly

caterpillar

hatch
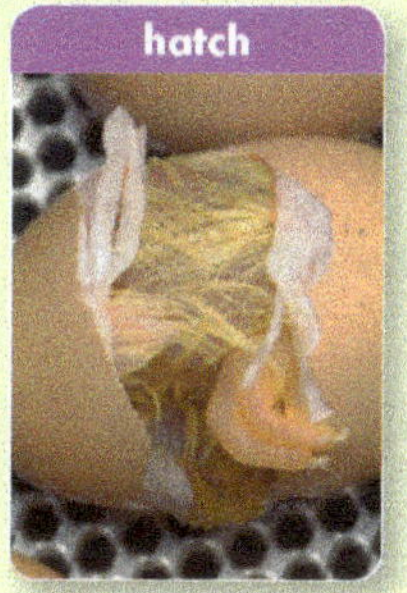

frog

tadpole

chrysalis

seed/seedling

fruit/apples

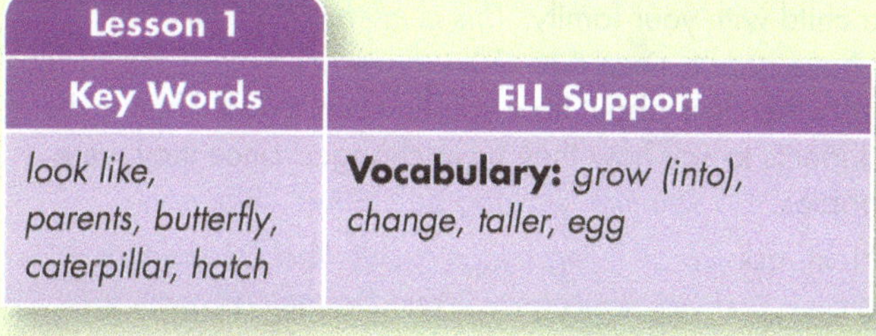

Lesson 1

Key Words	**ELL Support**
look like, parents, butterfly, caterpillar, hatch	**Vocabulary:** grow (into), change, taller, egg

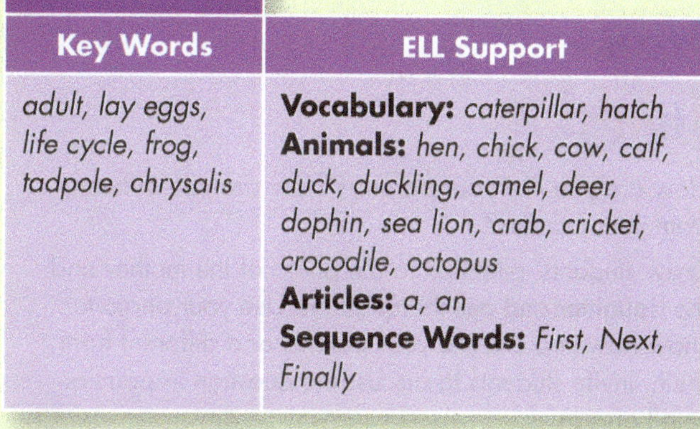

Lesson 2

Key Words	**ELL Support**
adult, lay eggs, life cycle, frog, tadpole, chrysalis	**Vocabulary:** caterpillar, hatch **Animals:** hen, chick, cow, calf, duck, duckling, camel, deer, dophin, sea lion, crab, cricket, crocodile, octopus **Articles:** a, an **Sequence Words:** First, Next, Finally

Lesson 3

Key Words	**ELL Support**
seedling, fruit	**Vocabulary:** seed, plant, root, stem, leaves, flower, apple seed, apple tree, grow into

Unit 4 — Plants and Animals

Unit Objectives

Lesson 1: Students will learn that living things grow and change.

Lesson 2: Students will learn how animals grow and change.

Lesson 3: Students will learn how plants grow and change.

Vocabulary: *living, nonliving, grow, change, tadpole, kitten, look like*

Introduce the Big Question

How do living things change as they grow?

Build Background Show a photo of yourself as a child with your family. *This is my family. This is me. I have grown since then. How have I changed?* Elicit ideas from the class about how you have changed. Then invite students to say how they have changed since they were babies. *Do you like growing up? Why? Why not?*

Next, ask *Do all living things grow? Yes! Do animals grow? Yes! Do plants grow? Yes! Do they all grow in the same way? No.*

Engage

Think!

How are you like your parents? How are you different from your parents?

Draw students' attention to the photo of the mother and the daughter and ask the question. Use your photo to show how you are like your parents and different from them. Invite students to discuss the question in pairs or small groups.

ELL Content Support

DNA carries instructions for our bodies to grow and change. DNA is made of genes. Genes determine how much babies look like their parents. Humans have many genes. Each person has two copies of each gene—one from each parent. Less than 1% of a person's genes are different from another person's genes. But, these small differences make us look different from other people.

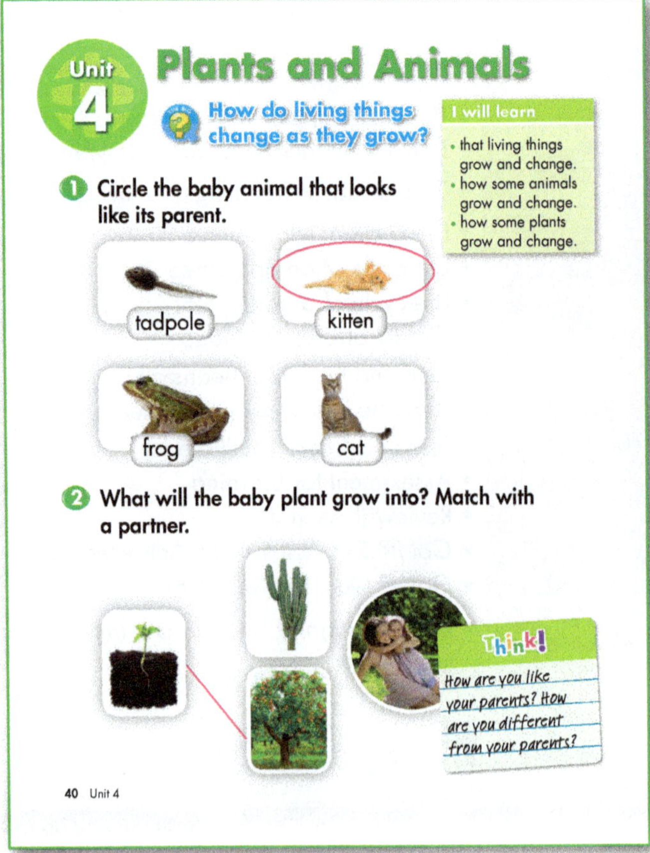

1 Circle the baby animal that looks like its parent.

Point to the pictures of the tadpole and the kitten. Elicit or tell students that they are babies. *What do they turn into when they grow? A frog and a cat. Which baby looks like its parent—the tadpole or the kitten?* Elicit the answer and what is similar about the kitten and the cat. Explain that tadpoles come out of eggs and they need time to transform into frogs.

2 What will the baby plant grow into? Match with a partner.

Have students look at the picture of the baby plant. Point out its leaves and root. *Will it grow into this plant (cactus) or this plant (tree)?* Elicit ideas and ask students to give reasons. (Possible answers. *A tree! It will grow into a tree.*)

Think! Again!

Revisit the question *How are you like your parents? How are you different from your parents?* Have the pairs or small groups share their answers with the class. Encourage students to bring in photos of their families to show to the class. Encourage students to talk about their siblings. Ask each student whether they look more like their mom or dad or a sibling.

Lesson 1

Do all young animals look like their parents?

Objective: Learn that living things grow and change.

Vocabulary: *grow, change, butterfly, caterpillar, look like*

Digital Resources: Flash Cards (*look like, butterfly, caterpillar*), Animal Cards (*cat and kitten, chicken and chick, cow and calf, dog and puppy, frog and tadpole, lion and lion cub*)

Unlock the Big Question

Write the following on the board: *I will learn about how living things change as they grow.*

Build Background Show the *look like* Flash Card to the students. *Does the baby chick look like its mom, the hen?* Elicit answers from the whole class and prompt students to explain or point to similarities and differences they can see. (Possible answers: similarities: *the same legs, similar eyes, the same beak, similar body*; differences: *different feathers, different color, different size wings, hen's wattle and comb*)

Explain

1 **Read. Match the babies with their parents.**

Read the paragraph for students. Invite them to match the babies to their parents and check with a partner. *Which is the calf's parent? The cow. Do they look alike? Yes. Which is the caterpillar's parent? The butterfly. Do they look alike? No.*

Call out other animals and ask students to say whether they look like their parents. For example, *Does a puppy look like its parent? Yes! Does a mouse look like its parent? Yes!*

ELL Vocabulary Support

Write *like* and *look like* on the board. Elicit an example with *like* and write it on the board, e.g., *I like butterflies.* Point out that we use *like* to talk about things we enjoy or don't enjoy. Elicit an example with *look like*, e.g., *My brother looks like my mom.* Point out that the words *look* and *like* go together in this sentence and mean *to be similar to someone or something.* You may also want to explain that we can use *am, is,* and *are like* instead of *look like.*

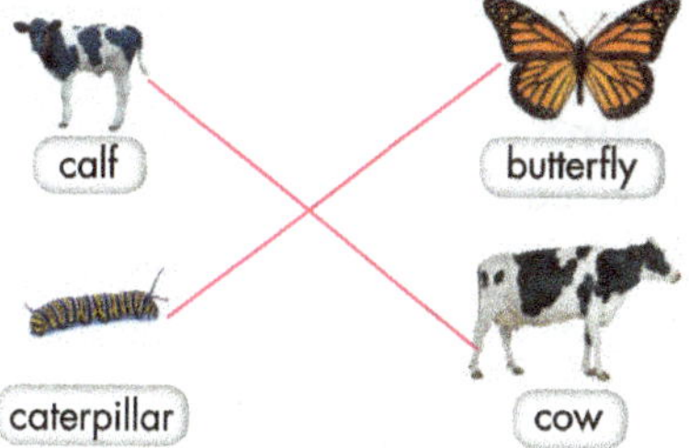

2 **How is the calf like the cow? Say with a partner.**

Have students read the question and discuss their ideas in pairs. Check answers as a class. (Possible answers: *It has four legs, two ears, a tail, and hair like its parent. It also has black and white hair.*)

Elaborate

Baby Animals

Display the Animal Cards and identify them with the students. *Which animals look like their parents most?* Ask students to look at the cards closely and decide as a class. *Which animals look like their parents least?* Elicit answers from the class. Put the students in pairs or small groups. Hand out an Animal Card to each pair or group. Ask students to draw a circle around the differences and discuss the differences and similarities. Monitor and provide help as necessary.

Think!

Why can't baby birds fly?

Activate prior knowledge about what animals need and different body coverings and body parts. Read the question aloud and elicit ideas. Guide students to conclude that baby birds need food from their mothers to grow bigger and stronger. They also need to grow feathers.

Do all young animals look like their parents?

Objective: Learn how people grow and change.

Vocabulary: *grow, change, taller*

Digital Resources: Flash Card (*look like*), Animal Card (*duck and duckling*, 1 for each student), *I Will Know...* Digital Activity

Materials: students' photos

Build Background Draw four stick figures on the board: a baby, a child, a teenager, and an adult. Name or elicit the different life stages and write them on the board under the pictures: *baby, child, teenager, adult. Which stage are you in? Child.*

Explain

3 **Look at the ducks. How are the young ducks like their parent? How are they different?**

Show the *look like* Flash Card and invite students to review the similarities and differences between the chick and the hen. Then draw students' attention to the picture of the ducks. Have students read and discuss their answers in pairs. Elicit answers from the class. Hand out a *duck* and *duckling* Animal Card to each student and invite them to color the animals.

4 **Read. What are some ways people grow and change? Talk in small groups.**

Read the paragraph for students. Draw their attention to the photos and elicit how old the children are. Invite students to think about how people grow and change and elicit answers from the class. Write students' ideas on the board. (Possible answers: *They get taller. They get stronger. They get bigger feet. They learn to walk. They learn to speak. They learn to read and write.*) Provide support as needed.

5 **Look at the picture. How does the boy change as he grows?**

Read the question aloud and draw students' attention to the boy in the picture. Elicit or say the answer. (*His teeth fall out, and new ones grow in.*) Prompt students to talk about their experiences with losing their teeth or their siblings losing their teeth.

I Will Know...

Have students do the *I Will Know...* Digital Activity.

---

3 Look at the ducks. How are the young ducks like their parent? How are they different?

4 Read. What are some ways people grow and change? Talk in small groups.

People Grow and Change

People are animals. People grow and change, too! They grow taller. They learn how to do new things, too.

5 Look at the picture. How does the boy change as he grows?

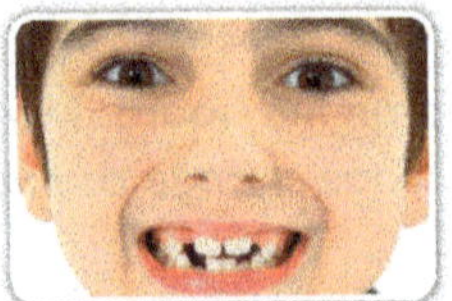

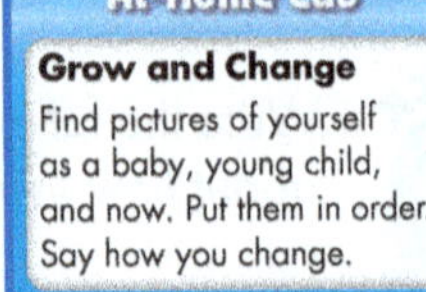
At-Home Lab

Grow and Change
Find pictures of yourself as a baby, young child, and now. Put them in order. Say how you change.

42   Unit 4   *I Will Know...*

Elaborate

At-Home Lab

Grow and Change

Assign the lab as homework. Ask students to find pictures of themselves at different ages and to bring them to class the next day. Invite students to show their pictures in small groups and explain how they have changed.

Most Important Change

Brainstorm with the class changes they have experienced, (Possible answers: *grow taller, learn to talk, walk, run, ride a bicycle,* etc.) Write a list on the board. Ask students to copy it in their notebooks in order of importance, with the item at the top being the most important change. Then invite students to compare their lists in pairs. Monitor and provide support as necessary.

Do all young animals look like their parents?

Objective: Learn that some animals hatch from eggs.

Vocabulary: *egg, hatch, grow into*

Digital Resources: Flash Card (*hatch*), variety of Animal Cards (animals that hatch, e.g., *crab, cricket, crocodile, eagle, fish, octopus*, and animals that don't hatch, e.g., *camel, deer, dolphin, sea lion*), *Lesson 1 Check* (print out 1 per student)

Materials: hard-boiled egg, photos from the Internet of different eggs (e.g., ostrich, shark, penguin, etc.)

Build Background Show the hard-boiled egg to the class. *What is this? Where does it come from? Do you eat eggs? What do you know about eggs?* Elicit answers from the students. Give the egg to a student to pass around the class. Invite students to study the egg and say what they like about it or want to know more about. Provide vocabulary support as necessary.

Explain

6 Read. Underline some animals that come from eggs.

Show the *hatch* Flash Card and ask students to describe what they see in the picture. Write *hatch* on the board and explain what it means (*to come out of an egg*). Invite students to say whether they have ever seen an animal hatch from an egg.

Read the paragraph and ask students to underline the animals that come from eggs. Check answers with the class. Then hold up the Animal Cards one by one and have students say whether the animal hatches from an egg. Provide support as needed.

7 Look at the pictures. What happens next? Number with a partner.

Draw students' attention to the pictures. Point out that the egg is the first stage in a chick's life. Ask them to number the other two pictures in order and check answers. Point out that the hen lays eggs so a full circle is formed: egg—chick—hen—egg.

8 Compare. How is the chick like its parent? How is it different? Talk in small groups.

Put the students into small groups. Read the questions aloud and encourage groups to discuss them. Monitor and provide support as necessary. Check answers as a class.

---

6 Read. Underline some animals that come from eggs.

Some Animals Hatch from Eggs

Many animals start as eggs.

Fish start as eggs.

Chickens start as eggs, too.

Chicks hatch from eggs.

Then they grow into chickens.

You can see the chick's feet and feathers.

7 Look at the pictures. What happens next? Number with a partner.

8 Compare. How is the chick like its parent? How is it different? Talk in small groups.

9 Do all chicken eggs grow into chickens? Talk with a partner.

---

9 Do all chicken eggs grow into chickens? Talk with a partner.

Show your hard-boiled egg again and ask *Is there a chick inside? No. Why? Do all chicken eggs grow into chickens?* Allow students to discuss in pairs. Guide students to understand that some chicken eggs don't grow into chickens.

Elaborate

What animal egg is it?

Show students pictures of different animal eggs, e.g., penguin, shark, snake, turtle, ostrich. Ask them to guess what animal lays each egg and to compare sizes. It does not matter if students guess correctly, only that they engage in comparing the eggs and animals.

Evaluate

Lesson 1 Check Assessment for Learning

Review the Key Words for Lesson 1 (see Student's Book page 41). Distribute the *Lesson 1 Check* and guide students as they complete it. Check answers as a class. Then ask students to grade their progress on the topic of how some animals look like their parents and others don't from 1 to 3: 3 = *I understand how some animals look like their parents*; 2 = *I need to study more*; 1 = *I need help!* Encourage students giving themselves a 1 or a 2 to say what they found difficult and what they need to study more.

How do some animals grow and change?

Objective: Learn about a frog's life cycle.

Vocabulary: *adult, lay eggs, life cycle, frog, tadpole*

Digital Resources: Flash Cards (*tadpole, caterpillar*)

Materials: magazine photos of a baby, a child, a teenager, and an adult (male or female)

Unlock the Big Question

Write the following on the board: *I will learn how some animals grow and change.*

Build Background Attach the photos to the board in random order. Then invite students to put them in order from the baby to the adult. Rearrange the photos so that they form a big circle. Add arrows from one photo to another starting with the baby and ending with the adult. *This circle is a life cycle.* Write *life cycle* on the board. *We're going to learn about an animal's life cycle! Which animal do you think it is?*

Explain

1 Read, look, and mark (✔) the adult animal.

Read the paragraph aloud for students. Explain *adult* by asking *Are you adults? No! Am I an adult? Yes!* Point to the pictures in the life cycle and invite volunteers to say what happens at each stage.

Then draw students' attention to the pictures on the right. Ask them to mark the correct adult animal. Check answers. Ask additional questions. *Do worms and fish hatch from eggs? Yes.*

ELL Vocabulary Support

You may wish to take the opportunity to go over the articles *a* and *an*. Write on the board: *I am a child. He is an adult.* Point out the difference between *a* and *an*. Then write the following sentence frames: *I am ___ person. I am ___ animal. It is ___ egg. It is ___ frog.* Invite students to say the articles that go in the blanks.

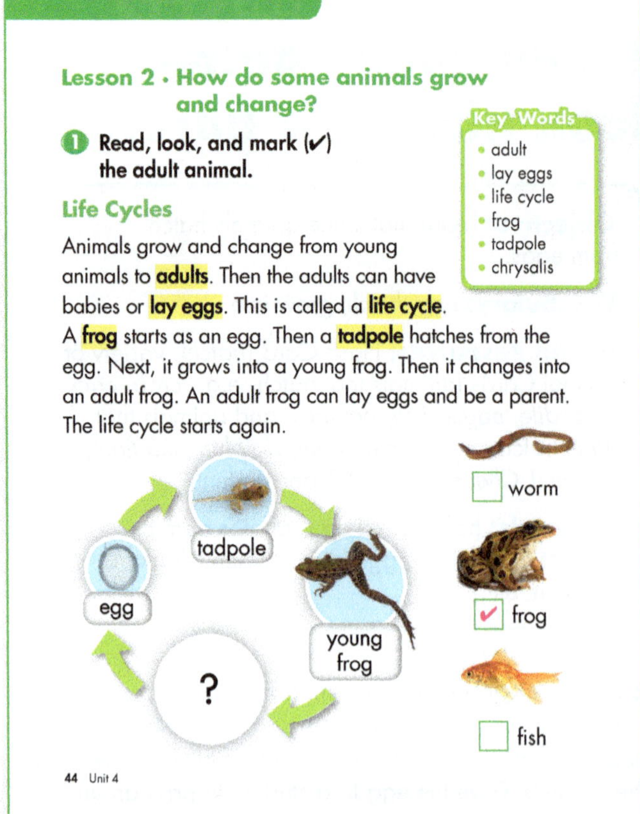

Elaborate

Worm and Fish Life Cycles

Have students choose the worm or fish and draw a life cycle for it. Monitor and make sure they include arrows to show the order of life stages. Help students label the pictures *egg, baby worm/fish, young worm/fish, adult worm/fish.* Invite students to show their pictures to the class.

Think!

Do frogs always complete their life cycles?

Activate prior knowledge by reminding students what they know about chicken eggs. Then ask the question and guide students to think about reasons a frog, worm, or fish might not be able to complete its life cycle. (Possible answer: *If it does not get food or water. Another animal can eat it.*) Discuss the same question for worms and fish.

How do some animals grow and change?

Objective: Learn about a frog's and a butterfly's life cycles.

Vocabulary: *adult, lay eggs, life cycle, frog, tadpole, chrysalis, caterpillar, hatch*

Digital Resources: Flash Cards (*tadpole, frog, caterpillar, chrysalis, butterfly*), *I Will Know…* Digital Activity

Materials: blue, green, and brown construction paper, foil circles, bubble wrap, black permanent marker, glue

Build Background Display the *tadpole* and *frog* Flash Cards on the board. Draw a question mark to the left of each Flash Card. *What comes before the tadpole? The egg! What comes before the adult frog? The young frog! Now we're going to learn about a butterfly's life cycle!*

Explain

2 **Look at the pictures. What happens next? Number with a partner.**

Read the question and draw students' attention to the pictures. Pair students and give them time to number the pictures in order. Check answers as a class.

3 **Read. How does a butterfly grow and change? Talk in small groups.**

What do you know about butterflies? Display the *caterpillar* and *butterfly* Flash Cards. Elicit what students know about the caterpillar from Lesson 1. Then display the *chrysalis* Flash Card. Invite volunteers to say what they think it is.

Read the paragraph aloud for students. Ask them to discuss the question in small groups. Then ask comprehension questions. *How does a butterfly start? As an egg. What comes out of the egg? A caterpillar. What does the caterpillar change into? A chrysalis. What does the chrysalis change into? A butterfly!*

4 **Read and trace. Say with a partner.**

Read the stages in a butterfly's life cycle. Have students trace the words. Then have pairs read a sentence each aloud. Monitor and help as necessary.

5 **Does a butterfly grow and change like a frog does? Talk as a class.**

Display the Flash Cards on the board. Ask the questions and start a class discussion. Provide support as needed.

---

2 Look at the pictures. What happens next? Number with a partner.

 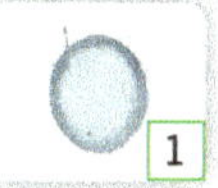 

3 Read. How does a butterfly grow and change? Talk in small groups.

The Life Cycle of a Butterfly

A butterfly starts as an egg. A caterpillar hatches from the egg. The caterpillar gets bigger. It changes into a **chrysalis**. A chrysalis changes into a butterfly. An adult butterfly can lay eggs. The cycle starts again.

4 Read and trace. Say with a partner.

A butterfly starts as an egg.

It changes into a caterpillar.

Then it changes into a chrysalis.

Finally, it grows into an adult butterfly!

5 Does a butterfly grow and change like a frog does? Talk as a class.

I Will Know… Unit 4 **45**

ELL Vocabulary Support

You may wish to teach sequence words to the class. Write the sequence words *First, Next,* and *Finally* on the board. Explain that these words help us describe stages. *First, a butterfly starts as an egg. Next, it… Finally, it…* Elicit the stages from the class.

Elaborate

Make a Frog's Life Cycle Poster

Cut out big circles of blue paper for each student. Explain that each circle is a pond. Hand out a small piece foil. Have students cut circles and decorate their pond. Then show them the bubble wrap. Cut out one bubble and draw a dot with the marker. *This is the egg.* Hand out pieces of bubble wrap for students to make eggs and stick them on the left-hand side of the pond. Then show students how to cut out tadpoles and young frogs from the brown paper. Help them stick the tadpoles and young frogs in the middle of the pond. Next, give students a piece of green paper to draw and cut out a frog or frogs. Ask students to stick them on the right-hand side of the pond. Monitor and help as necessary.

Finally, as a class, review the steps in a frog's life cycle: egg, tadpole, young frog, adult frog. Have students point to each stage as it is mentioned.

I Will Know…

Have students do the *I Will Know…* Digital Activity.

How do some animals grow and change?

> **Objective:** Review how animals grow and change.
>
> **Vocabulary:** *adult, lay eggs, tadpole*
>
> **Digital Resources:** Flash Cards (*tadpole, frog*), 1 set of Animal Cards per small group (*panda, monkey, dog and puppy, cat and kitten, eagle, firefly, pelican, kangaroo, frog, rabbit*), Lesson 2 Check (print out 1 per student)

Build Background Activate prior knowledge by asking students to describe a frog's life cycle in pairs. Have students use their artwork from the previous lesson.

Explain

6 **Circle *T* (true) or *F* (false). Check with a partner.**

Read the sentences and ask the students to circle *T* or *F* on their own. Have students compare their answers in pairs and then check as a class. Ask them to correct the false sentence. (1. *Animals do change.*)

7 **Draw a tadpole and an adult frog. How are they different? Show and tell with a partner.**

Display the Flash Cards and ask the students to draw a tadpole and an adult frog in their books. Monitor and provide support as necessary. Have the same pairs compare their drawings and describe the differences between a tadpole and a frog. Elicit differences from the whole class. (Possible answers: *A tadpole is small. A frog's legs are big. Its head and mouth are big, too.*)

8 **What are two animals that come from eggs? Say as a class.**

Display the Animal Cards on the board. In small groups, ask students to look at the cards and decide which animals come from eggs. Elicit ideas.

9 **How does a puppy change? Say as a class.**

Display the *dog* and *puppy* Animal Card. Ask the question and elicit ideas. Review other animals that have babies rather than lay eggs, e.g., monkeys, pandas, cats, cows, etc.

Elaborate

Animal Groups

Divide the class into two groups: Animals That Lay Eggs and Animals That Have Babies. Call out animals. *Shark!* The corresponding group stands up. Continue with different animals. Provide support as needed.

---

6 Circle *T* (true) or *F* (false). Check with a partner.

a) Animals don't change. T /(F)

b) A young frog is called a tadpole. (T)/ F

c) An adult animal can have babies or lay eggs. (T)/ F

7 Draw a tadpole and an adult frog. How are they different? Show and tell with a partner.

tadpole   adult frog

8 What are two animals that come from eggs? Say as a class.

9 How does a puppy grow and change? Say as a class.

At-Home Lab

Grow and Change
Tell your family how a butterfly or frog changes as it grows.

46   Unit 4   Lesson 2 Check

At-Home Lab

Grow and Change

Assign the lab as homework. Encourage students to use their artwork or a picture from the Student's Book to explain the butterfly or frog life cycle. Have them report back to the class.

Evaluate

Lesson 2 Check Assessment for Learning

Review the Key Words for Lesson 2 (see Student's Book page 44). Distribute the *Lesson 2 Check* and guide students as they complete it. Check answers as a class. Then ask students to grade their progress on the topic of how some animals grow and change from 1 to 3: 3 = *I understand how some animals grow and change*; 2 = *I need to study more*; 1 = *I need help!* Encourage students giving themselves a 1 or a 2 to say what they found difficult and what they need to study more.

How do some plants grow and change?

Objective: Learn how some plants grow and change.

Vocabulary: *seed, plant, seedling, root, stem, leaves, flower*

Digital Resources: Flash Card (*seed/seedling*), *Let's Explore!* Digital Lab

Materials: plant, ideally with flowers, a handful of sunflower seeds in the shell, permanent marker; 1 per student: beans, plastic bag, wet paper towel

Unlock the Big Question

Write the following text on the board: *I will learn how plants grow and change.*

Build Background Display the plant. Point to different parts and invite students to name them (*stem, leaves, flower, petal, roots*).

Explore

Let's Explore! Lab   What things are living?

Objective: Observe how a bean plant grows.

Digital Resources: *Let's Explore!* Digital Lab, *Let's Explore! Activity Card* (print out 1 per student) (*Optional:* Show the Digital Lab.)

- Show the beans and elicit what they are. *Are they living or nonliving? Living.* Then show the plastic bag and the paper towel. *We are going to start growing beans!* Invite students to think about how.

- Hand out the materials. Ask students to put a small piece of paper towel in the bag, place the bean on top, and cover it with the rest of the paper towel. Monitor and write the students' names on the bags with a permanent marker.

- For the next two to three days, have students add a little bit of water to the paper towel so that it stays damp. The bean plant should sprout, and, over the next two weeks, students will see the bean plant grow.

- Have students work in pairs to complete the *Activity Card.* Monitor and provide support as needed.

Explain

1 **Read. What are four parts of a sunflower plant? Underline.**

Draw students' attention to the picture of the sunflower. *What is this plant called?* Read the

---

Lesson 3 · How do some plants grow and change?

1 **Read. What are four parts of a sunflower plant? Underline.**

Key Words
- seedling
- fruit

The Life Cycle of a Plant

Plants grow and change, too. Most plants grow from a seed. A sunflower starts life as a seed. The seed has food for the young plant. The young plant is called a seedling. The sunflower grows roots, a stem, and leaves. It gets taller. It grows into an adult sunflower plant. The plant makes flowers. And the flowers make seeds. The life cycle can start again.

2 **How does a sunflower seed grow into a sunflower plant? Talk with a partner.**

sunflower seeds

3 **Does a daisy plant grow like a sunflower? Talk with a partner.**

Think!
Do all plants have flowers?

Let's Explore! Lab   Unit 4   47

paragraph aloud. Have students underline the four parts of the sunflower plant as you read. Have the class say the four parts. Allow students to touch and describe the sunflower seeds. Then crack a few open and look at the seed inside.

2 **How does a sunflower seed grow into a sunflower plant? Talk with partner.**

Have pairs discuss the question. Then check answers as a class. (Possible answers: *The seed grows into a seedling. When the plant is an adult, it starts making flowers.*)

3 **Does a daisy plant grow like a sunflower? Talk with a partner.**

Have students find a picture of a daisy in their books (page 35, 36, or 38). In pairs, students discuss the question. Elicit answers from the class. Guide students to understand that both sunflowers and daisies grow from seeds and have a similar life cycle.

Think!

Do all plants have flowers?

Have students discuss freely. Invite volunteers to share any experiences they may have had with plants that do not have flowers. Accept all logical answers. Students may mention or describe pine trees, ferns, or moss as plants that do not have flowers.

How do some plants grow and change?

Objective: Learn about the life cycle of an apple tree.

Vocabulary: *apple seeds, adult apple tree, seedling, fruit*

Digital Resources: Flash Card (*fruit/apples*), *I Will Know…* Digital Activity

Materials: an apple cut in half, seeds from different types of fruit (e.g., watermelon, peach, strawberry)

Build Background Display the Flash Card. *What are they? Do you eat them? Are they good for you? What other fruit do you eat? Do they come from a plant?*

Explain

4 **Read. Look at the pictures. Draw arrows to show the life cycle of the apple tree.**

Read the paragraph aloud to students. Invite volunteers to say one stage in the apple tree's life cycle at a time.

Ask the students to draw the arrows. Monitor and provide help as necessary.

5 **Draw a fruit you know. Does it have seeds? What do the seeds look like?**

Hold the apple halves together and show the fruit to the class. Ask *Does it have seeds? Where are they? What do they look like?* Elicit answers. Then separate the two halves and show the students. *Can you eat the seeds? No.*

Ask the students to think of a different fruit. Elicit ideas and write them on the board. Put the students in pairs to answer the questions about each fruit.

Give students time to draw a fruit and its seeds. Invite volunteers to show their drawings and talk about the seeds.

I Will Know…
Have students do the *I Will Know…* Digital Activity.

---

4 **Read. Look at the pictures. Draw arrows to show the life cycle of the apple tree.**

The Life Cycle of an Apple Tree

An apple tree starts as a seed. Then it changes into a seedling. Then it grows into an adult apple tree. An adult apple tree grows flowers. Some of the flowers make **fruit** with seeds. The apple seeds can grow into apple seedlings.

5 Draw a fruit you know. Does it have seeds? What do the seeds look like?

 I Will Know...

ELL Content Support

All types of fruit have seeds that can grow into baby plants. Some fruit (e.g., cherries, peaches, plums) contain a pit or stone that has the seed inside. Other types of fruit (e.g., melons, watermelons, pineapples) have lots of seeds inside them, while other types (e.g., strawberries, raspberries) have lots of seeds on the outside. Ensure students understand that not all seeds are edible.

Think!

Why are there different types of fruit seeds?

Show the seeds you have brought and identify them with the students' help. *Why are there different types of fruit seeds?* Guide students to conclude that different kinds of fruit seeds grow into different kinds of fruit.

How do some plants grow and change?

Objective: Learn about bean plants.

Vocabulary: *bean seed, seedling, adult bean plant, grow into*

Digital Resources: Flash Card (*seed/seedling*), *Lesson 3 Check* (print out 1 per student), *Got it? 60-Second Video* (optional: Lesson 2 *Let's Explore!* Lab)

Materials: dried beans

Build Background Hand out a dried bean to each student. *What are they? Bean seeds. Do you eat beans? How do they grow?* (You may wish to show the Lesson 2 *Let's Explore!* Lab in class for exercise 6.)

Explain

6 **Read. Draw a bean, a bean seedling, and an adult bean plant. Show and tell.**

Read the paragraph along with students and display the *seed/seedling* Flash Card. Ask comprehension questions. *How does a bean start? As a seed. What does it grow into? A seedling. What happens next? It grows into an adult bean plant.* Review the stages on the Flash Card. Invite volunteers to show their drawings.

7 **Think of a plant or animal. Draw it as a young plant or animal. Draw it as an adult. Show and tell. How does the plant or animal change?**

Read the questions and give students time to think of a plant and animal. Invite volunteers to show their drawings and explain how their chosen plant or animal changes. Provide help as necessary.

Elaborate

Jack and the Giant Beanstalk

Write the story title on the board and elicit whether students know it. Tell the story with the students' help and focus on how the bean seed grows into a giant bean plant.

- Jack and his mother are poor. They live off the milk they get from their cow.
- They decide to sell their cow at the market, but Jack exchanges it for magic seeds from a strange old man.
- His mother gets angry and throws the bean seeds out the window. The next day, a beanstalk starts to grow. It grows really fast and gets really big.
- Jack climbs up the beanstalk and meets a giant who has a magic hen; it lays golden eggs.

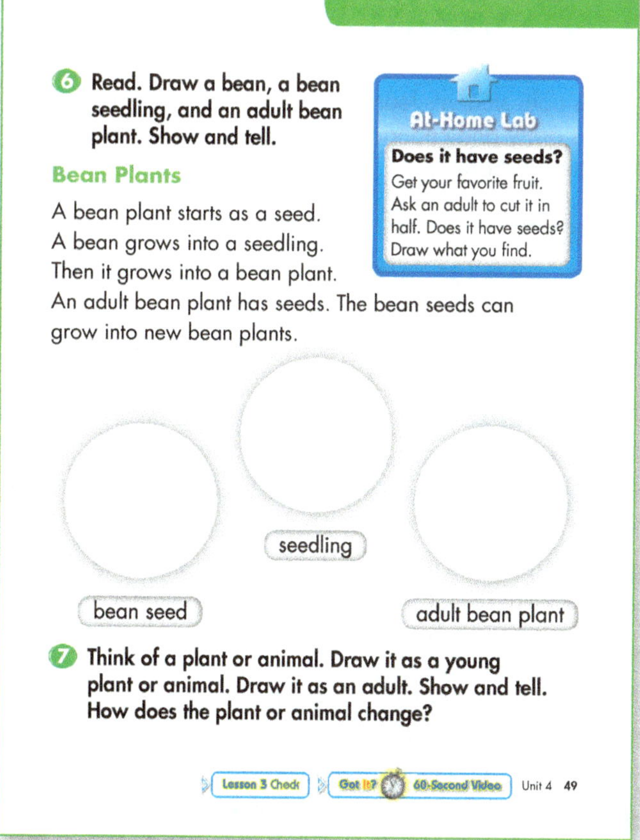

- Jack steals the hen, and the giant chases him down the beanstalk. Jack cuts the beanstalk down.

Evaluate

Lesson 3 Check   Assessment for Learning

Review the Key Words for Lesson 3 (see Student's Book page 47). Distribute the *Lesson 3 Check* and guide students as they complete it. Check answers as a class. Then ask students to grade their progress on the topic of how some plants grow from 1 to 3: 3 = *I understand how some plants grow and change;* 2 = *I need to study more;* 1 = *I need help!* Encourage students giving themselves a 1 or 2 to describe what they found difficult and what they need to study more.

Play the *Got it? 60-Second Video* to review the unit material.

Let's Investigate!

In this unit, students learn about some ways animals and plants grow and change. In this lab, they will observe how a butterfly changes.

Let's Investigate! Lab — **How does a butterfly change?**

Objective: Students will observe how a butterfly grows and changes.

Materials: sheets of white paper, caterpillars, butterfly home, crayons.

Digital Resources: *Let's Investigate!* Digital Lab, *Let's Investigate! Activity Card* (1 per student)

- Display the picture of the butterfly home and explain what it is. Help students remember the stages in a butterfly's life cycle. *How does a butterfly start its life? What comes out of the egg? What does a caterpillar turn into? What does a chrysalis turn into?*

- Next, display the caterpillars on a table. Hand out the sheets of paper and ask students to divide their sheets of paper into three parts. Have students look at the caterpillars and draw them on the first part of the sheet.

- Place the caterpillars inside the butterfly home with leaves to feed on. Within two weeks, the caterpillars should turn into chrysalises. Ask students to observe them and draw a chrysalis on the second part of their sheets.

- Within a maximum of 30 days, the chrysalises should transform into butterflies. Ask students to observe and draw a butterfly on the last part of their sheets.

- If the season is appropriate, take the students and the butterfly home outside. With the students' help, free the butterflies and watch them as they fly away.

Teacher Time-Saving Option: Show the *Let's Investigate!* Digital Lab as an alternative to the hands-on lab activity.

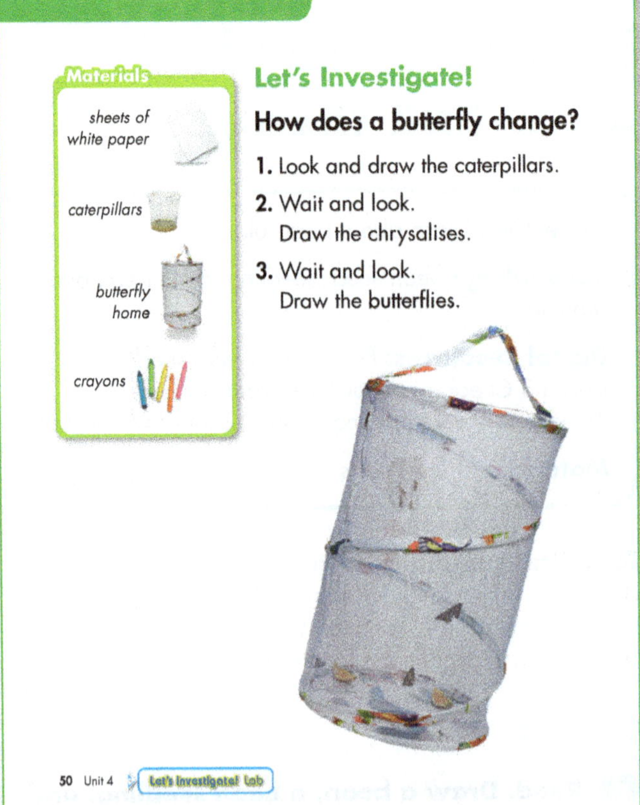

Class Project: Seeds Poster

Materials: different types of safe seeds (flowers, fruit, vegetables, nuts, etc.) with corresponding pictures of the flowers, fruit, vegetables, and nuts from the Internet or magazines, tape, construction paper (1 per pair or small group), crayons or colored pencils

Let's create a seeds poster! Display the seeds on a table together with the corresponding pictures. Ask students to look at the seeds and the pictures and talk about them. Then have pairs or groups make life cycle posters for one of the seeds. Help students label the seeds and pictures. Display the posters in class.

Unlock the Big Question

Have students refer to the Big Question on the Unit Opener page. In pairs have them recall what they have learned about how animals and plants grow and change. Have students answer questions 3 and 4 on the on the *Activity Card*.

Unit 4 Review

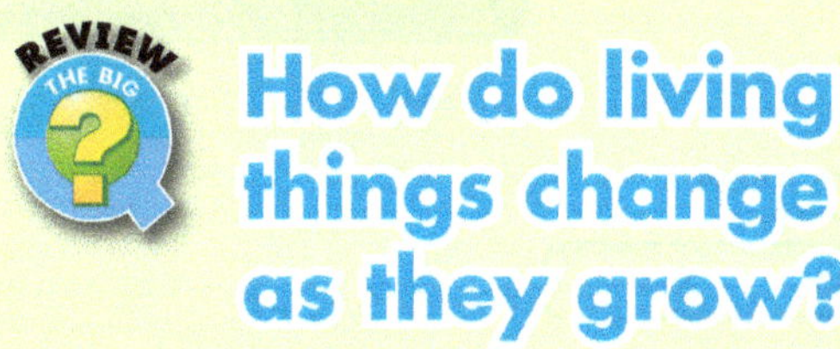

How do living things change as they grow?

Digital Resources: Print out 1 of each per student: *Got it? Self Assessment, Got it? Quiz*

Evaluate

Strategies for Targeted Review

The following are strategies for providing targeted review for students if they encounter challenges with the content.

Lesson 1 Do all young animals look like their parents?

Question 1

If... students are having difficulty marking the sentences true or false, then... review the photos of the plants and animals in Lesson 1. Then ask students to correct the false sentences.

Lesson 2 How do some animals grow and change?

Question 2

If... students are having difficulty understanding the steps, then... ask them to find photos of an egg, a tadpole, and an adult frog in Lesson 2.

Lesson 3 How do some plants grow and change?

Question 3

If... students are having difficulty identifying the pictures, then... invite students to review the lesson and the pictures.

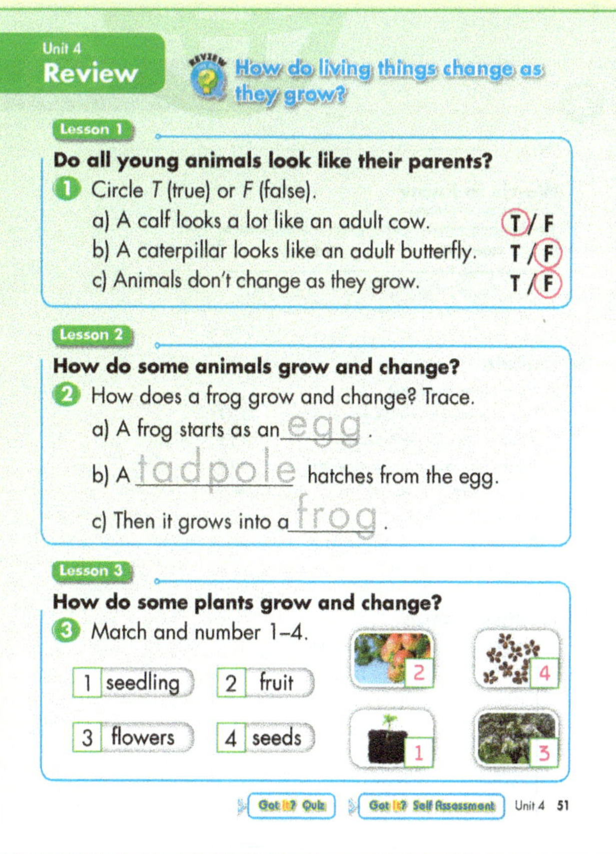

ELL Language Support

Before students start working on the Review activities, have them read each question aloud along with you.

Got it? Self Assessment

Immediately after students have completed the Review activities, distribute a *Got it? Self Assessment* to each student. Have students complete the *Stop! Wait!* and *Go!* statements for each lesson, allowing them to look back through the lesson material if necessary.

Got it? Quiz

Distribute a Unit 4 *Got it? Quiz* to each student. Quizzes may be used for assessing students' understanding of unit concepts as well as for grading purposes.

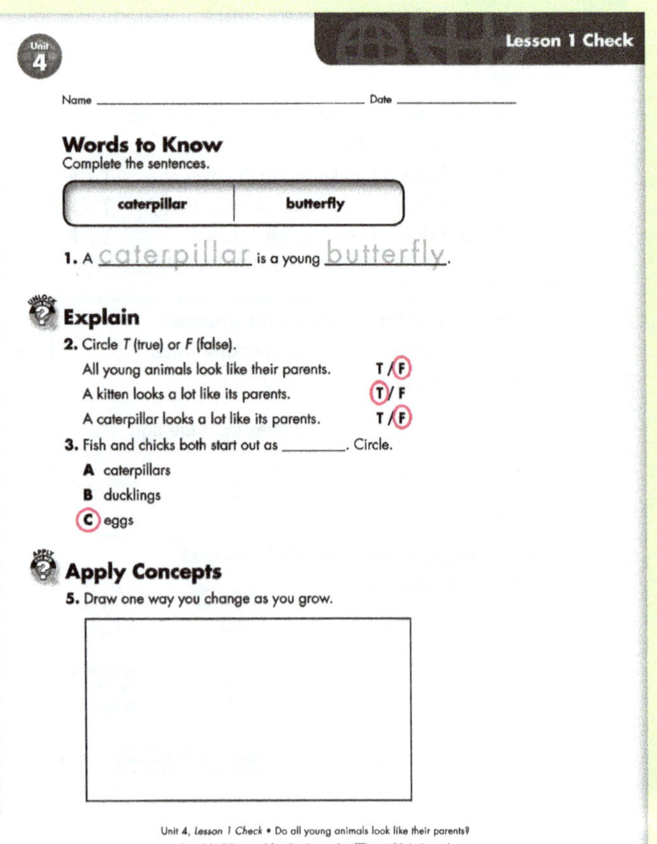

Name _____________________ Date _____________________

Words to Know
Complete the sentences.

| caterpillar | butterfly |

1. A caterpillar is a young butterfly.

Explain

2. Circle *T* (true) or *F* (false).

All young animals look like their parents.   T / **F**

A kitten looks a lot like its parents.   **T** / F

A caterpillar looks a lot like its parents.   T / **F**

3. Fish and chicks both start out as _________. Circle.

A caterpillars

B ducklings

C eggs

Apply Concepts

5. Draw one way you change as you grow.

Unit 4, Lesson 1 Check • Do all young animals look like their parents?
Copyright © Pearson Education, Inc., or its affiliates. All Rights Reserved.

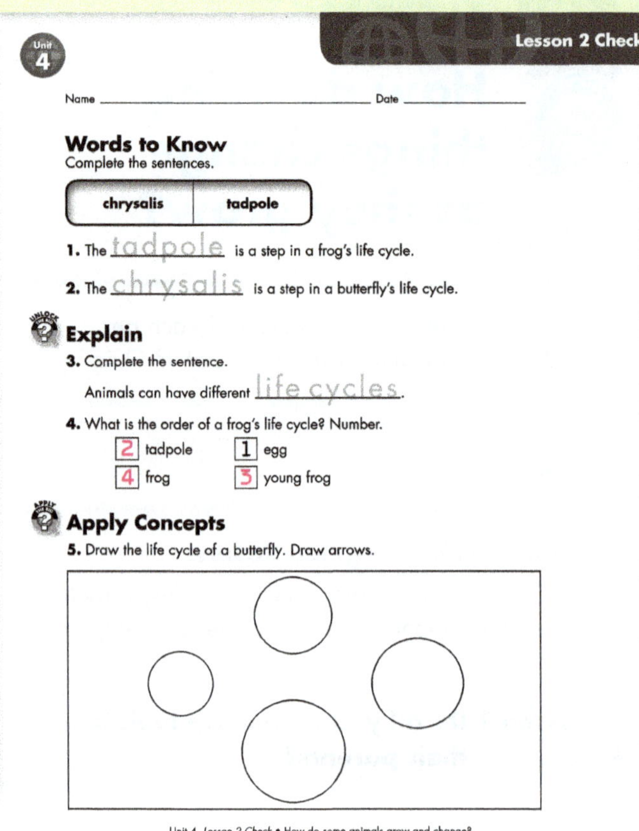

Name _____________________ Date _____________________

Words to Know
Complete the sentences.

| chrysalis | tadpole |

1. The tadpole is a step in a frog's life cycle.

2. The chrysalis is a step in a butterfly's life cycle.

Explain

3. Complete the sentence.

Animals can have different life cycles.

4. What is the order of a frog's life cycle? Number.

[2] tadpole   [1] egg

[4] frog   [3] young frog

Apply Concepts

5. Draw the life cycle of a butterfly. Draw arrows.

Unit 4, Lesson 2 Check • How do some animals grow and change?
Copyright © Pearson Education, Inc., or its affiliates. All Rights Reserved.

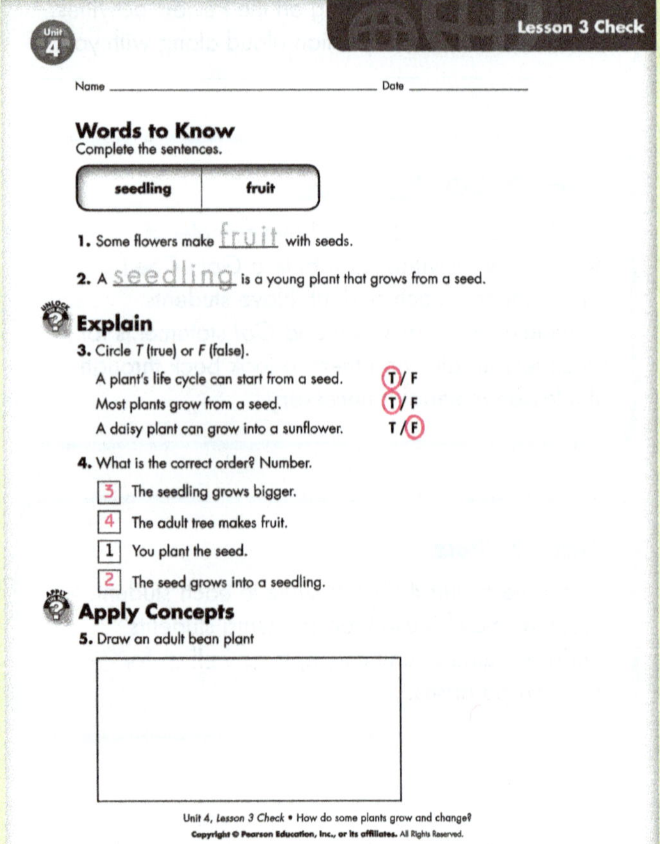

Name _____________________ Date _____________________

Words to Know
Complete the sentences.

| seedling | fruit |

1. Some flowers make fruit with seeds.

2. A seedling is a young plant that grows from a seed.

Explain

3. Circle *T* (true) or *F* (false).

A plant's life cycle can start from a seed.   **T** / F

Most plants grow from a seed.   **T** / F

A daisy plant can grow into a sunflower.   T / **F**

4. What is the correct order? Number.

[3] The seedling grows bigger.

[4] The adult tree makes fruit.

[1] You plant the seed.

[2] The seed grows into a seedling.

Apply Concepts

5. Draw an adult bean plant

Unit 4, Lesson 3 Check • How do some plants grow and change?
Copyright © Pearson Education, Inc., or its affiliates. All Rights Reserved.

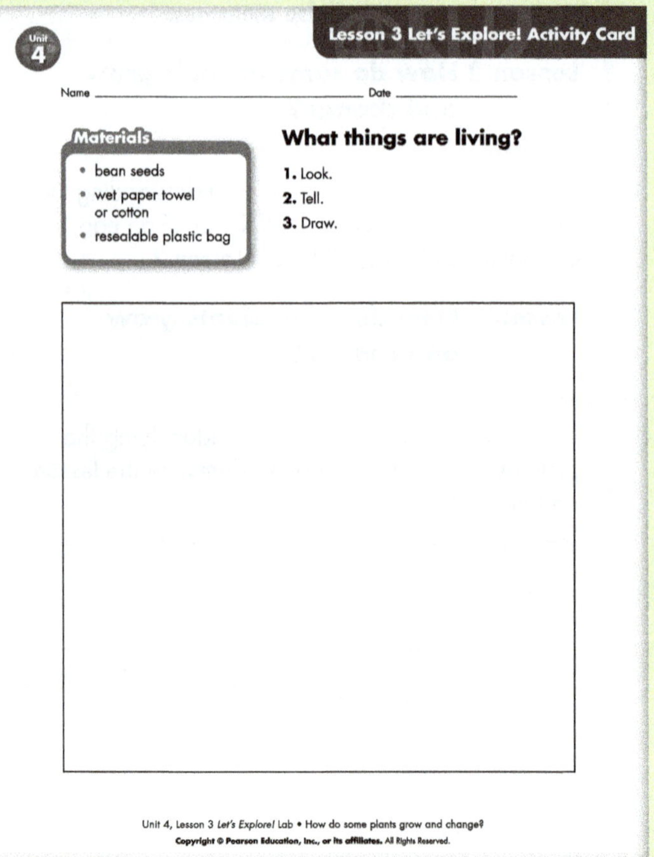

Name _____________________ Date _____________________

Materials
- bean seeds
- wet paper towel or cotton
- resealable plastic bag

What things are living?

1. Look.

2. Tell.

3. Draw.

Unit 4, Lesson 3 Let's Explore! Lab • How do some plants grow and change?
Copyright © Pearson Education, Inc., or its affiliates. All Rights Reserved.

T51a   Unit 4 • Digital Resources and Photocopiables

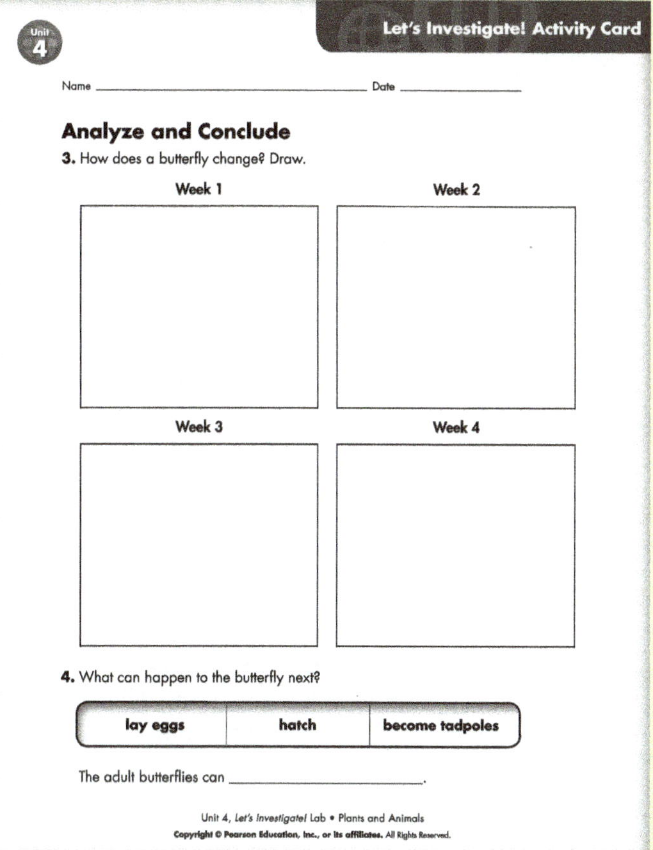

Name _______________________ Date _______________

Analyze and Conclude

3. How does a butterfly change? Draw.

Week 1	Week 2

Week 3	Week 4

4. What can happen to the butterfly next?

lay eggs	hatch	become tadpoles

The adult butterflies can ___________________.

Unit 4, *Let's Investigate! Lab* • Plants and Animals
Copyright © Pearson Education, Inc., or its affiliates. All Rights Reserved.

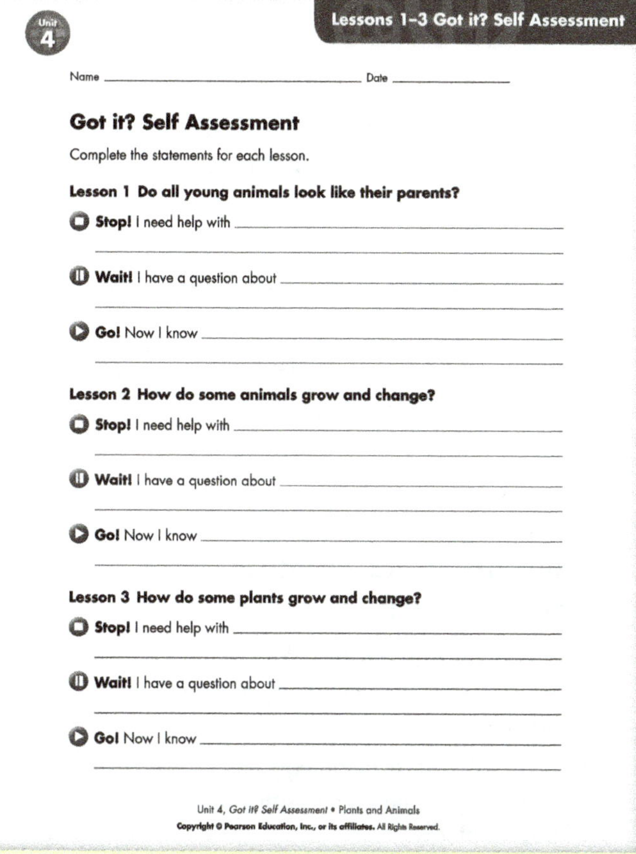

Name _______________________ Date _______________

Got it? Self Assessment

Complete the statements for each lesson.

Lesson 1 Do all young animals look like their parents?

Stop! I need help with _______________________

Wait! I have a question about _______________________

Go! Now I know _______________________

Lesson 2 How do some animals grow and change?

Stop! I need help with _______________________

Wait! I have a question about _______________________

Go! Now I know _______________________

Lesson 3 How do some plants grow and change?

Stop! I need help with _______________________

Wait! I have a question about _______________________

Go! Now I know _______________________

Unit 4, *Got it? Self Assessment* • Plants and Animals
Copyright © Pearson Education, Inc., or its affiliates. All Rights Reserved.

Name _______________________ Date _______________

Got it? Quiz

1. What baby animal looks like its parents? Circle.
(A) a calf
B a tadpole
C a caterpillar

2. What baby animal does NOT look like its parents? Circle.
A a calf
B a kitten
(C) a caterpillar

3. What can happen as people grow and change? Circle.
(A) They can learn new things.
B They can get smaller.
C They can hatch from eggs.

4. What animals hatch from eggs? Circle.
A puppies
B kittens
(C) chicks

Unit 4, *Got it? Quiz* • Plants and Animals
Copyright © Pearson Education, Inc., or its affiliates. All Rights Reserved.

Name _______________________ Date _______________

5. How many steps are there in a frog's life cycle? Circle.
A two
B three
(C) four

6. An adult butterfly can _________. Circle.
(A) lay eggs
B jump
C whistle

7. A young plant that grows from a seed is called _________. Circle.
(A) a seedling
B a duckling
C a daisy

8. A bean plant starts as _________. Circle.
A an egg
(B) a seed
C a trunk

Unit 4, *Got it? Quiz* • Plants and Animals
Copyright © Pearson Education, Inc., or its affiliates. All Rights Reserved.

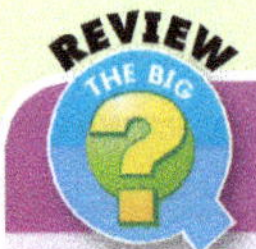

Unit 4 Study Guide

How do living things change as they grow?

Lesson 1
Do all young animals look like their parents?

- Kittens and calves look a lot like their parents.
- Caterpillars and tadpoles look different from their parents.

Lesson 2
How do some animals grow and change?

- A frog starts as an egg, then grows into a tadpole, then a young frog, and then an adult frog.
- A butterfly starts as an egg, then grows into a caterpillar, then a chrysalis, and then an adult butterfly.

Lesson 3
How do some plants grow and change?

- A sunflower starts life as a seed, then grows into a seedling, and then an adult sunflower.
- Some plants have flowers that make seeds.

Review the Big Question

How do living things change as they grow?

Have students use what they have learned from the unit to answer the question in their own words.

How has your answer to the Big Question changed since the beginning of the unit? What are some things you learned that caused your answer to change?

Make a Concept Map

Draw on the board a concept map like the one shown on this page. With the students, talk through the key ideas from this unit. Invite different students to point to the ideas on the board, miming as possible.

Unit 4 Concept Map

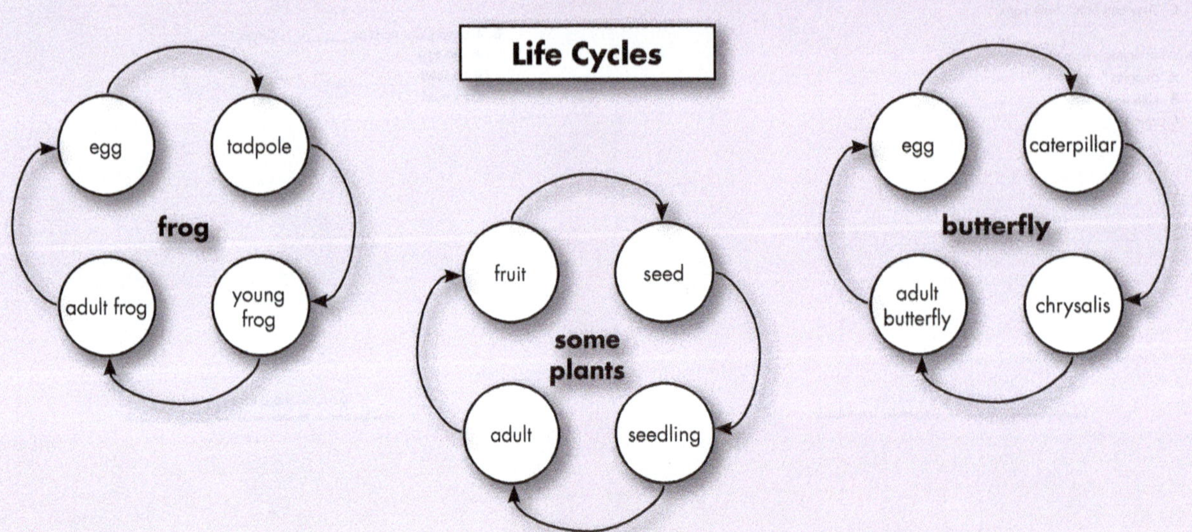

Students can make a concept map to help review the Big Question.

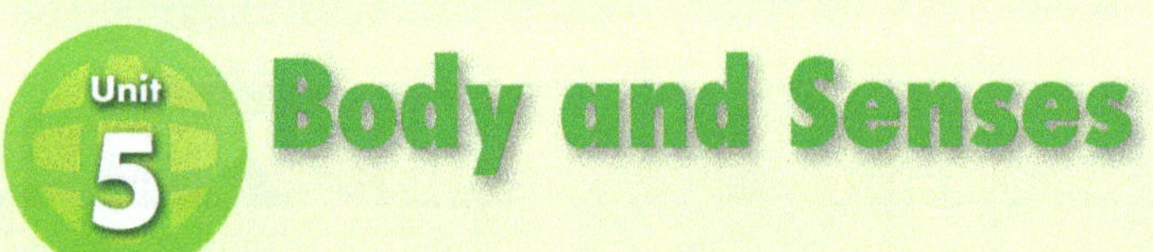 Body and Senses 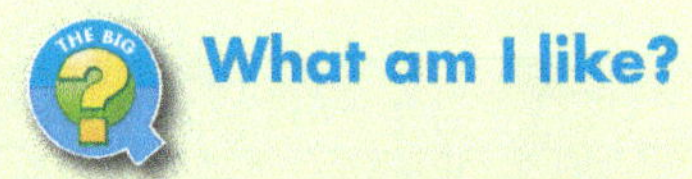 What am I like?

Lesson Plan

Unit Opener & Lesson 1 What are my senses?			
	Activity	**Pages**	**Time**
Engage	• Unit Opener: Think! *Can we hear everything dogs hear?*	SB p. 52	10 min
	• Unit Opener: Identify the five senses and talk about what we see.	SB p. 52	10 min
	• Unit Opener: Think about what our body parts tell us about the soup.	SB p. 52	10 min
	• Unit Opener: Think about what your body needs.	SB p. 52	10 min
	• Think! *Why does an ambulance make a loud sound?*	TB p. 55	10 min
	• Think! *What is hot and what is cold?*	TB p. 56	10 min
	• Think! *Why are my sneakers and feet smelly?*	TB p. 57	10 min
	• Think! *What sense do you use a lot?*	SB p. 58	10 min
Explain	• Which parts of our bodies we use for each of our senses	SB p. 53	30 min
	• What we can see	SB p. 54	30 min
	• What sounds we can hear	SB p. 55	30 min
	• What we can touch and feel	SB p. 56	30 min
	• What we can smell	SB p. 57	30 min
	• What we can taste	SB p. 58	30 min
Elaborate	• Senses Game	TB p. 53	20 min
	• Rainbow: The Order of Colors	TB p. 54	20 min
	• Animal Sounds	TB p. 55	10 min
	• Flash Lab: Things You Can See and Hear	SB p. 55	10 min
	• What's in the pillowcase?	TB p. 56	20 min
	• Smell It, Name It	TB p. 57	10 min
	• Taste without smell?	TB p. 58	20 min
Evaluate	• *Lesson 1 Check* (Active Teach)	TB p. 63a	10 min
	• Assessment for Learning	TB p. 58	10 min
	• Review (Lesson 1)	SB p. 63	10 min
	• *Got it? Self Assessment* (Active Teach)	TB p. 63b	10 min
	• *Got it? Quiz* (Active Teach)	TB p. 63b	10 min

Lesson 2 What does my body need?			
	Activity	**Pages**	**Time**
Engage	• Think! *Do you need to eat candy?*	SB p. 59	10 min
	• Think! *Why do I sweat?*	TB p. 60	10 min
	• Think! *Do our senses help us stay healthy and safe?*	TB p. 62	10 min
Explore	• Digital Lab: *Is it healthy or unhealthy?* (ActiveTeach)	TB p. 59	30 min
Explain	• Food that your body needs	SB p. 59	30 min
	• Exercise and why it's good for us	SB p. 60	30 min
	• Water and why we need it	SB p. 61	30 min
	• Why we need water and shelter	SB p. 62	30 min
	• *Got it? 60-Second Video* (ActiveTeach)	TB p. 62	10 min
Elaborate	• Let's exercise!	TB p. 60	30 min
	• Animal Shelters	TB p. 61	20 min
Evaluate	• *Lesson 2 Check* (ActiveTeach)	TB p. 63a	10 min
	• Assessment for Learning	TB p. 63	10 min
	• Review (Lesson 2)	SB p. 63	10 min
	• *Got it? Self Assessment* (ActiveTeach)	TB p. 63b	10 min
	• *Got it? Quiz* (ActiveTeach)	TB p. 63b	10 min
Lab	• *Let's Investigate! How many points can you feel?* (ActiveTeach)	TB p. 62	30 min

Flash Cards

see

hear

touch/feel

taste

smell

healthy

exercise

sleep

shelter

Lesson 1

Key Words	ELL Support
see, hear, touch, taste, smell, feel, skin, tongue	**Vocabulary:** *color, size, shape, bark, sing, loud, soft, rough, smooth, hot, cold, sweet, salty, sour* **Questions:** *What are you like? What is he/she like?* **Can/Can't for Ability:** *It's sunny. I can see the trees. It's dark. I can't see anything.* **Colors and Shapes**

Lesson 2

Key Words	ELL Support
healthy, energy, exercise, sleep, shelter	**Vocabulary:** *water, grow, run, play, think, bone, muscle, strong, remember, learn, safe, warm*

Unit 5 — Body and Senses

Unit Objectives

Lesson 1: Students will learn about the five senses.

Lesson 2: Students will learn what their bodies need.

Vocabulary: *see, hear, touch, taste, smell, feel, skin, tongue, healthy, energy, exercise, sleep, shelter*

Introduce the Big Question

What am I like?

Build Background *In this unit, we are going to learn about our senses and our bodies.* Remind students that they learned about the five senses in Unit 1. Ask them to turn to Student's Book page 4 and look at the pictures. *What body parts can you see in exercise 2?* (*eye, ear, mouth/tongue, hand, nose*) Write the body parts on the board and say them aloud with the students.

ELL Vocabulary Support

Read the Big Question. Write on the board: *What are you like? What is he/she like?* Explain that we use these questions to ask what a person looks like and what a person is like. Elicit examples. *I'm tall and funny. He is smart and shy.*

Engage

Think!

Can we hear everything dogs hear?

Point to the picture of the dog. *What animal is it? What is it like? Does it have small or big ears?* Allow groups to freely discuss the *Think!* question. Monitor and provide support as necessary.

1 **Look at the picture. What do you see? Say with a partner.**

Point to the picture of the hot air balloon and teach the words. Ask students to describe what they see. (Possible answers: *It's yellow, pink, blue, red, and orange. It's round. It flies in the air. The sky is blue. There aren't any clouds.*) *How can you describe the picture? What body part do you use?* (*My eyes.*)

2 **Look at the pictures. What body parts can tell you about the soup? Talk in pairs.**

Invite students to identify the body parts pictured. Then draw their attention to the picture of the soup. *What's the soup like? Is it hot? Is it salty?*

Is it colorful? Does it smell nice? Elicit answers from the students and encourage them to explain which body parts they would use to describe different aspects of the soup. (Possible answers: *I can use my eyes to say what it looks like. I can use my tongue and hands to tell if it's hot or cold. I can use my tongue to taste it. I can use my nose to smell it.*)

3 **What does your body need? Say as a class.**

Read the question aloud and elicit what each picture is. Then have pairs discuss what their bodies need. If necessary, point out that *need* here means *something we can't live without*. *Does your body need food? Does it need toys?*

Elicit ideas from the whole class and ask students to explain them. Ask further questions. *Does your body need sunshine? Does it need music? Does it need heat?* It does not matter at this point if students fully understand the difference between needs and wants, only that they start engaging in the topic.

Think! Again!

Have students revisit the question *Can we hear everything dogs hear?* Ask questions *Can dogs hear us when we talk to them? Yes! Can we hear when dogs bark? Yes! Do you think we can hear everything dogs can?* Accept all logical answers.

What are my senses?

Objective: Learn about our five senses.

Vocabulary: *see, hear, touch, taste, smell*

Digital Resources: Flash Cards (*senses, observe, see, hear, touch/feel, taste, smell*)

Materials: sensory objects, e.g., soap with strong fragrance, a smooth stone, a bell, apple slices (1 for each student plus 1 additional one)

Unlock the Big Question

Write the following on the board: *I will learn about my five senses.*

Build Background Display the *senses* Flash Card. Elicit the body parts and invite volunteers to write the words on the board: *ears, tongue, eyes, hands, nose. How do you take care of your ears? How do you take care of your eyes?* Elicit ideas. (Possible answers: *I wash my ears. I cover them if it's too loud. I put on safety goggles in science class.*)

Explain

1 Read. Look at the pictures. Label the pictures with a partner.

Display the *observe* Flash Card and invite students to describe what they see. *Who is she? What is she doing?* (*She's a scientist. She's observing something. Here, she's using her eyes to observe.*) Read the paragraph aloud for students and have them label the pictures with the highlighted words. Check answers by displaying the Flash Cards (*see, hear, touch/feel, taste, smell*). Say the five senses for the students to repeat, and correct pronunciation.

2 Look at the pictures. Match the words to the sentences.

Read the sentences one by one and have students underline the verbs (*see, hear, touch, taste, smell*). Draw students' attention to the body parts and have them match the words to complete the sentences. Invite volunteers to read the sentences aloud.

ELL Vocabulary Support

You may wish to take the opportunity to explain the difference between *can* and *can't* for ability. Say *It's sunny. I can see trees outside.* and write *can see* on the board. Say *It's dark. I can't see anything outside.* and write *can't see. Close your eyes. Can you see anything? No. Hold your nose. Can you smell anything? No.*

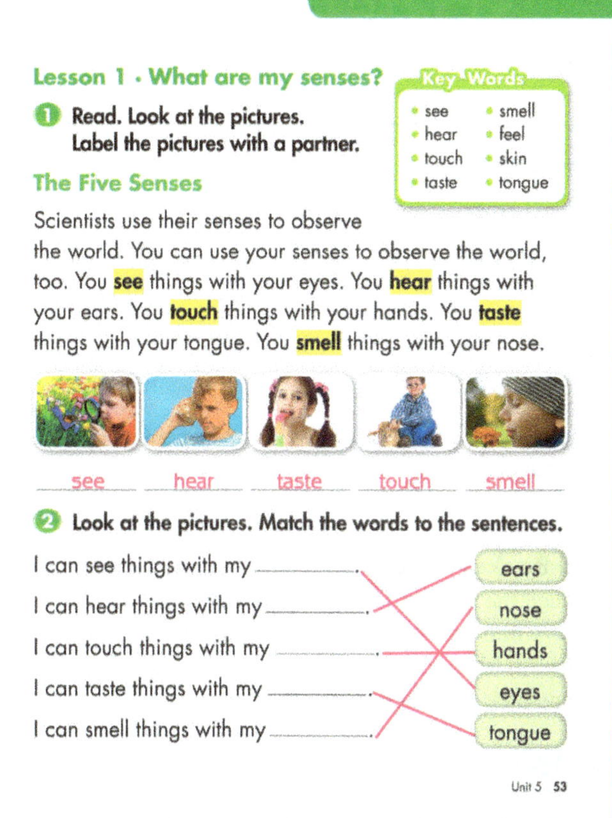

Elaborate

Senses Game

Divide the class into five groups, Eyes, Ears, Mouth/Tongue, Hands, and Nose. Have students stand in their groups. First, pass around the objects one by one and allow students to observe them. Encourage them to use their senses to do so. Explain that they should only use their sense of taste to eat one apple slice each.

Next, display the materials one at a time and have the appropriate groups stand up and point to the body part. For example, when you hold up the soap, Eyes, Noses, and Hands should stand up. When you ring the bell, Eyes, Ears, and Hands should stand up.

At the end of the exercise, ask students to discuss which groups stood up the most.

Think!

How do you use your senses in class?

Ask the question and have students discuss it in pairs. Monitor and provide help as necessary. Elicit and accept all logical answers from the class. (Possible answers: *We use our senses like scientists. We use our senses to observe things. We use our senses to learn about the world around us!*)

What are my senses?

Objective: Learn what kinds of things we can see.

Vocabulary: *see, color, size, shape*

Digital Resources: Flash Cards (*see, hear, touch/ feel, taste, smell*), Animal Cards (*camel, chipmunk, crab, crocodile, dolphin, giraffe, kangaroo, monkey, octopus, rabbit, tiger, bear*), *I Will Know…* Digital Activity

Materials: sheets of paper, colored pencils or crayons

Build Background Show the five senses Flash Cards and elicit the words. *Today we are going to learn about what we can see.* Next, display the *see* Flash Card and brainstorm ideas about what students can see. Then have students describe the picture of the girl.

Explain

3 **Read. What kinds of things can you see? Talk in small groups. Underline.**

Read the question and the paragraph aloud. Have students discuss it in groups. Ask them to underline the answers. Check answers as a class. (Answers: *colors, sizes, shapes, big or small, a circle or a square.*) Invite students to describe the picture of the girl again. Read the caption aloud.

ELL Vocabulary Support

Review vocabulary for colors (*brown, white, black, red, blue, green, yellow, orange,* etc.) and shapes (*circle, square, rectangle, triangle, diamond,* etc.). Point to colors and draw shapes to elicit the vocabulary words and write them on the board.

4 **Look at the picture. What colors can you see? Say with a partner.**

Ask students to look at the picture. *What is it? A rainbow. Where do you see it? In the sky. When? When it's raining and sunny.* Have pairs name the colors in the rainbow. Check answers as a class.

ELL Content Support

Rainbows can appear when there are drops of water in the atmosphere. Light hits the drops of water and makes the colors we see in rainbows. We can see rainbows when it's raining. We can also see them when it's misty and foggy. But, there has to be sunlight hitting the water droplets for rainbows to be visible.

---

3 Read. What kinds of things can you see? Talk in small groups. Underline.

See

You can see <u>colors</u>, <u>sizes</u>, and <u>shapes</u>. You can see what color something is. You can see how big or small something is. You can tell if something is a circle or a square.

The girl's umbrella has white circles!

4 Look at the picture. What colors can you see? Say with a partner.

5 What animal is big? What animal is small? Point and say with a partner.

6 Look at the picture. What things can you see? Say as a class.

---

5 **What animal is big? What animal is small? Point and say with a partner.**

Point to the picture of the elephants. *What animals are they? Do they look like each other?* Elicit ideas. Then have students point to the small elephant and the big elephant. *Which is the baby? The small elephant. Which is the parent? The big one.* Then hold up two Animal Cards. Ask students to identify the small and big animals. Continue with different pairs of Animal Cards. *You're observing like scientists!*

6 **Look at the picture. What things can you see? Say as a class.**

Have the students describe the picture. Invite volunteers to call out things they see. (Possible answers: *pink flowers, red petals, a plant with flowers, green leaves, white clouds, a blue sky*)

Elaborate

Rainbow: The Order of Colors

How many colors are there in a rainbow? Six. Draw a rainbow on the board. Invite students to say the colors in order. Write them on the board: *red, orange, yellow, green, blue, purple.* Distribute the materials. Have students draw and color their own rainbows. Invite volunteers to show their drawings to the class.

I Will Know…

Have students do the *I Will Know…* Digital Activity.

What are my senses?

> **Objective:** Learn what kinds of things we can hear.
>
> **Vocabulary:** *hear, bark, sing, loud, soft, bee, cow, worm*
>
> **Digital Resources:** Flash Card (*hear*)
>
> **Materials:** Animal Cards (*cat and kitten, chicken and chick, cow and calf, dog and puppy, duck and duckling, frog and tadpole, horse, lion and lion cub, monkey, mosquito, parrot, snake*)

Build Background Display the *cat and kitten* and *dog and puppy* Animal Cards on the board. Point to one card at a time. Elicit or say the animal for the students to repeat. Then put the *hear* Flash Card on the board. *What's this sense? What sound does a cat make?* Invite volunteers to meow like a cat. *What sound does a dog make?* Invite students to bark like a dog. Leave the cards on the board.

Explain

7 **Read. What are two things you can hear? Talk with a partner.**

Read the paragraph with the students. Explain *loud* and *soft* by clapping your hands loudly and softly. Have pairs think of two things they can hear. Students can refer to the text for examples. Share answers as a class.

ELL Vocabulary Support

You may wish to take the opportunity to explain the words *bark*, *sing*, and *meow*. Write the words on the board and have the students identify which animal they refer to (*dog, bird, cat*).

8 **What makes a loud sound? What makes a soft sound? Point.**

Draw students' attention to the pictures. *What is this?* An ambulance. *What is this?* A bird. Ask the questions and elicit or provide the answers.

What other things or animals make a loud sound? What other things or animals make a soft sound? Elicit answers from the whole class.

9 **Look at the pictures. What animals can you hear? Say as a class.**

Draw students' attention to the pictures and elicit the animals. Put students in pairs to answer the question and share answers as a class. (Answers: *cow, bee*)

---

7 Read. What are two things you can hear? Talk with a partner.

Hear

You can hear a dog bark or a bird sing. You can hear loud sounds and soft sounds. A dog can make a loud sound. A bee can make a soft sound.

> **Flash Lab**
>
> **Things You Can See and Hear**
> Walk around the classroom. Say four things you can see. Say two things you can hear.

8 What makes a loud sound? What makes a soft sound? Point.

9 Look at the pictures. What animals can you hear? Say as a class.

Think!

Why does an ambulance make a loud sound?

Ask the question and have students discuss their ideas in small groups. (Possible answer: *So that other drivers can hear it and get out of the way.*) Invite students to think about other loud sounds that are used in emergencies. (Possible answers: *police cars, fire engines, fire alarm, subway doors closing,* etc.) Provide support as needed.

Elaborate

Animal Sounds

Display the rest of the Animal Cards on the board. Invite volunteers to mime the sound each animal makes. Then ask *Which animal makes a loud sound? Which animal makes a soft sound? Which animal makes a funny sound?* Accept all logical answers.

⚡ Flash Lab

Read the instructions and ask *Which senses are you going to use?* See and hear. Give students a minute to walk around the classroom and notice four things they can see and two they can hear. Ask them to sit down and share what they saw and heard in small groups. Share ideas as a class.

What are my senses?

> **Objective:** Learn about what things feel like.
>
> **Vocabulary:** *feel, skin, hands, touch, rough, smooth, hot, cold*
>
> **Digital Resources:** Flash Card (*touch/feel*)
>
> **Materials:** a piece of scratchy burlap, a scarf, rough items (sandpaper, rough and smooth stone, string), smooth items (leaf, apple, spoon, flower petals), pillowcase, thick sock, and various small items (e.g., eraser, ruler, different types of fruit, cotton ball, sponge, feather, toys)

Build Background Display the *touch/feel* Flash Card. Explain the difference between the two verbs. Touch the scarf as you say *Touch.* Then say *It feels nice.* Ask the students to touch something near them. Invite volunteers to say what they are touching.

Explain

10 Read. Does your desk feel smooth or rough?

Read the paragraph with the students. Draw their attention to the pictures and explain the difference between *rough* and *smooth. Touch your desk. Is it smooth or rough?* Elicit answers from the class. Display the items you have brought on a table. Invite students to come up and touch the items carefully. Then ask them to sit down. Hold up one item at a time. *Is it smooth or rough?* Allow students to touch each item again if necessary.

11 Look at the pictures. What feels soft? Circle.

Ask the students to look at the pictures and have them circle the items they think feel soft. Check answers as a whole class. (Answers: *teddy bear, dog's fur*)

12 Look at the pictures again. What feels rough? Talk with a partner.

Pass around the piece of burlap for students to feel. *Does it feel soft? No!* Then ask the students to look at the pictures in exercise 11 again. *What feels rough?* Elicit answers from the class. (Answer: *The material.*)

ELL Content Support

Skin is the largest organ of the human body. It protects everything in our bodies: our bones, our muscles, and all our other organs. Through our skin, we can feel touches and temperature.

10 Read. Does your desk feel smooth or rough?

Touch and Feel

You can use your hands to touch things. You **feel** things with your **skin.** You can feel when something is rough or smooth. Some rocks are rough. Some rocks are smooth. You can feel what is hot and what is cold. Be careful. Don't touch anything hot!

11 Look at the pictures. What feels soft? Circle.

12 Look at the pictures again. What feels rough? Talk with a partner.

56   Unit 5

Think!

What is hot and what is cold?

Divide the class into small groups and ask them to think of as many hot and cold items as they can. Give them one or two minutes. Then elicit ideas from the class. (Possible answers: hot: *pot, stove, oven, fire,* etc.; cold: *ice, snow, ice cream, fridge,* etc.)

Elaborate

What's in the pillowcase?

Keep the items you brought secret and place one in the pillowcase. Walk around the class and invite a student to put their hand into the pillowcase and touch the item. Then ask them to say what it is and take it out so that everyone can see. Provide vocabulary support as necessary. Repeat with different items and students. As they touch the items, ask *Is it soft? Is it small?*

Continue with items that are harder to identify. Then ask the students to wear the sock on their hand and continue the procedure. *Is it easier or harder to identify the items? Harder!*

What are my senses?

Objective: Learn about things we can smell.

Vocabulary: *smell good or bad, flowers, garbage, perfume, sun, bell, pony*

Digital Resources: Flash Card (*smell*)

Materials: different items to smell (soap, mosquito repellent, orange peel, coffee, garlic, pencil shavings, cotton ball soaked in vinegar), cloth bag

Build Background Display the *smell* Flash Card and invite students to describe what they see. (Possible answers: *There's a boy. He's smelling a flower. The flower is orange. He's wearing a hat.*) *Does he like the smell of the flower? Yes. Do you like smelling flowers?*

Explain

13 Read. What smells good? Mark (✔). What smells bad? Mark (✗).

Read the sentences and elicit or explain what each picture shows. Have students mark the pictures and check answers in pairs. Share answers as a whole class. Invite students to think of more things that smell good or bad. Write them on the board and then hold a class vote on the worst smell.

14 What smells good? Talk in small groups. Circle.

Read the question and say the words under the pictures for the students to repeat. Put students in small groups to decide what smells good. Have them circle the pictures. Check answers as a class.

15 What are some things you can hear, see, and touch? Talk with a partner.

Point to the three pictures and elicit the words from the students. If necessary, explain that a pony is a small horse. Read the question aloud and have the students answer it in pairs. Check answers with the class.

What other things can you see, hear, and touch? In the same pairs, invite students to think of other things. (Possible answers: *dog, cat, monkey, car, alarm clock*, etc.)

13 Read. What smells good? Mark (✔). What smells bad? Mark (✗).

Smell

You use your nose to smell. Many flowers smell good. Sometimes garbage smells bad.

14 What smells good? Talk in small groups. Circle.

15 What are some things you can hear, see, and touch? Talk with a partner.

Elaborate

Smell It, Name It

Keep all the items inside a bag or box. Invite a few volunteers to stand at the front of the class. Ask them to close their eyes. Then walk past each volunteer and hold one of the items close to their nose. Ask them to smell it and make sure they are not looking. Put the item in the bag again. Ask the student to open their eyes and guess what they smelled. Check and repeat with different volunteers and items. Provide vocabulary support as needed.

Set all the items on a table. Discuss as a class which items were easy to identify by their smell.

Think!

Why are my sneakers and feet smelly?

Do your sneakers and feet smell bad sometimes? When? Why? Discuss with students that people's shoes and feet often smell bad in spring and summer when it's hot. Their feet sweat and lots of bacteria start growing on them. *What can you do about this problem?* Elicit or explain that the best way to deal with smelly feet is to wash them with soap and dry them well. It also helps to change socks daily, wash your shoes, and wear sandals in summer.

Lesson 1

What are my senses?

> **Objective:** Learn more about how and what we taste.
>
> **Vocabulary:** *taste, sweet, salty, sour, tongue, ice cream, popsicle, potato chips, lime*
>
> **Digital Resources:** Flash Cards (*see, hear, touch/ feel, taste, smell*) *Lesson 1 Check* (print out 1 per student)
>
> **Materials:** candy, salty cracker, a lime, a knife (teacher use), different kinds of juice (e.g., apple, mango, orange), milk, water, plastic cups

Build Background Invite students to call out foods and drinks they have had today. Write words on the board and ask *Do you like the taste of (bread)?* Students stand up for *Yes* or stay seated for *No.* Display the *taste* Flash Card. *What is the girl eating? A popsicle. Do you like the taste of popsicles?* Provide support as needed.

Explain

16 **Read. Look at the pictures. What tastes sweet? What tastes salty? Say with a partner.**

Show the candy. *What is this? Candy. It's sweet.* Then show the crackers. *What are these? Crackers. They're salty.* Write *sweet* and *salty* on the board. Allow students to taste a piece of each. Show the Flash Card again. *Is the ice cream sweet or salty? Sweet.* Read the paragraph aloud. Write *sour* on the board and show the lime. *Is this sweet? No. Is it salty? No. It's sour.* Have pairs point to each picture and say the taste. Check answers as a class.

17 **Look at the pictures again. Circle what you can taste and smell.**

Hold up the lime. *Can you smell it?* Pass it around for the students to smell. *Yes.* Cut it in small pieces for students to taste. *Can you taste it? Yes. What does it taste like? Sour.* Have pairs decide whether they can taste and smell the foods pictured. Check answers as a class. Accept all logical answers. (Possible answers: *You can taste ice cream. You can taste and smell the potato chips.*)

18 **Look at the pictures. What things can you see or hear? What things can you smell or taste? What things can you feel? Talk as a class.**

Elicit what is pictured. *Can you see and hear the steak? We can see it, but we can't hear it. Can you smell it and taste it? Yes. Can you feel it? Yes.* Continue with questions about the moon and the alarm clock. Provide support as needed.

16 **Read. Look at the pictures. What tastes sweet? What tastes salty? Say with a partner.**

Taste

You taste things with your **tongue**. You can taste sweet things like ice cream. You can taste salty things like potato chips. You can taste sour things, too. Limes taste sour.

17 Look at the pictures again. Circle what you can taste and smell.

18 Look at the pictures. What things can you see or hear? What things can you smell or taste? What things can you feel? Talk as a class.

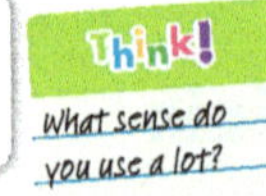

58 Unit 5   Lesson 1 Check

Think!

What sense do you use a lot?
Display the five senses Flash Cards. Students call out the words. Read the question and have students discuss it. Accept all logical answers. Explain that we don't always depend on our sense of seeing. *When we are in a dark room, our senses of hearing and smelling become stronger.*

Elaborate

Taste without smell?

Our noses help us to taste foods. They can help us even more than our tongues. Make a chart on the board labeled *With Smell* and *Without Smell.* Have volunteers close their eyes and taste different liquids while holding their noses. Invite them to guess what each liquid is. Add a check (✔) if they can guess or an ✘ if not. Repeat the procedure with volunteers not holding their noses to complete the second column. Then have students use the chart to discuss how the sense of smell helped them taste.

Evaluate

Lesson 1 Check Assessment for Learning

Review the Key Words for Lesson 1 (see Student's Book page 53). Distribute the *Lesson 1 Check* and guide students as they complete it. Check answers as a class. Then ask students to grade their progress on the topic of our senses from 1 to 3: 3 = *I understand our senses;* 2 = *I need to study more;* 1 = *I need help!* Encourage students giving themselves a 1 or a 2 to say what they found difficult and what they need to study more.

T58   Unit 5 • Body and Senses: What am I like?

Lesson 2

What does my body need?

> **Objective:** Learn about what our bodies need.
>
> **Vocabulary:** *healthy, energy, food, exercise, water, sleep, shelter, grow, run, play, think*
>
> **Digital Resources:** Flash Cards (*living/nonliving, grow, healthy*), *Let's Explore!* Digital Lab

Unlock the Big Question

Write the following text on the board: *I will learn about things that my body needs to stay healthy.*

Build Background Display the *living/nonliving* Flash Card. Ask students to remember what they learned about living and nonliving things in Unit 3. Allow them to look back at pages 28–39 of the Student's Book. Then display the *grow* Flash Card. *Do living things grow? Yes. Do nonliving things grow? No. What do living things need to grow properly?* Elicit ideas but don't give feedback.

Explore

Let's Explore! Lab   Is it healthy or unhealthy?

Objective: Learn what types of food are healthy or unhealthy.

Digital Resources: *Let's Explore!* Digital Lab, *Let's Explore! Activity Card* (1 per student)

- Spread the pictures of different types of food and drinks on a table. Ask the students to gather around the table and identify the food and drinks. Provide vocabulary support as necessary.

- Invite volunteers to pick up pictures of healthy and unhealthy food and drinks. Ask the class to say if they agree or not.

- Show the Digital Lab and invite groups to do the activity. Remind students they can look at the pictures for ideas as they go through the activity.

- Have students work in pairs to complete the *Activity Card.*

Explain

1 Read. Say three things you need with a partner.

Read the paragraph aloud. Explain what *exercise* means in this context (*running, jumping, playing*) and the meaning of *shelter* (*a place where you are safe*, e.g., a house). Put the students in pairs to say three things they need. Elicit answers from the class and discuss why each thing is important.

---

1 Read. Say three things you need with a partner.

Living things have needs. You are a living thing. You need food, exercise, water, sleep, and shelter.

2 Read. What does food give you? Underline.

Food

You need food to stay **healthy**. Food gives you **energy**. Energy helps you grow. It helps you do things like run and play and think in school.

3 Why do you need food? Talk in small groups.

---

2 Read. What does food give you? Underline.

Ask the question and have the students read the paragraph. Ask them to underline the things that food gives us. Check answers as a class. Explain the meanings of *healthy* and *energy* using the *healthy* Flash Card.

3 Why do you need food? Talk in small groups.

Have groups discuss the question. Monitor and provide support as necessary. Write answers on the board. (Possible answers: *It gives us energy. It helps us to grow and stay healthy.*)

Think!

Do you need to eat candy?

Ask the question and have students discuss their ideas in small groups. Discuss the question again as a class. (Possible answers: *We don't need to eat candy. We can eat other food. Eating a little bit of candy isn't too unhealthy.*) Invite students to think about how much candy they eat each day or week.

> **ELL Content Support**
>
> According to research, children between four and ten years old should have no more than six teaspoons of sugar each day. Many people don't realize that one can of soda pop usually has seven teaspoons of sugar and one muffin usually has five teaspoons of sugar!

What does my body need?

Objective: Learn about how exercise and water help your body.

Vocabulary: *exercise, healthy, bones, muscles, strong, water, run, play, drink*

Digital Resources: Flash Cards (*healthy, exercise*), *I Will Know…* Digital Activity

Build Background Activate prior knowledge by asking students what animals need. Then ask *What do you need to stay healthy?* Display the *healthy* Flash Card and elicit what the top left picture shows. Point to the top right and bottom left pictures. *Today we are going to learn about exercise and water. Why do we need them?*

Explain

4 **Read. How does exercise help your body? Say with a partner.**

Read the paragraph aloud for students. Ask each student to touch their upper arm and notice that it's soft. Explain that the soft part is the muscle. Then ask them to press harder to feel the bone. Elicit what they know about muscles and bones. *How does exercise help your body?* Have the students answer in pairs and as a class.

ELL Content Support

The human body has more than 600 skeletal muscles. Muscles help our blood move through our bodies. We can control many of our muscles (for example, our arm muscles), but there are some muscles we can't control (for example, our heart). A human body has 206 bones. More than half of our bones are in our hands (54) and our feet (52). Bones make up our skeleton and help us move. They also protect the organs that are inside our bodies.

5 **What are some ways you can exercise? Talk in small groups. Show.**

Draw students' attention to the pictures. Ask the question and have them discuss in small groups. Share answers as a class. Invite volunteers to stand up and show you and the class how they exercise.

6 **Read. Look at the picture. Is the boy doing something to keep his body healthy?**

Have the students look at the picture on the lower right of the page. *What is the boy doing? Drinking. What is he drinking? Water. Is he doing something*

---

4 **Read. How does exercise help your body? Say with a partner.**

Exercise

You need exercise to stay healthy. **Exercise** keeps your bones and muscles healthy and strong.

5 **What are some ways you can exercise? Talk in small groups. Show.**

6 **Read. Look at the picture. Is the boy doing something to keep his body healthy?**

Water

You need water, too. About half of your body is made of water! You lose water when you run and play or when it's hot outside. Then you need to drink more water.

60 Unit 5 I Will Know…

---

healthy? Yes. Read the paragraph with the students. *Why is water good for you? A lot of your body is made of water!*

Elaborate

Let's exercise!

Have students pretend they are in PE class and do a few different types of exercise (stretch arms and legs, run in place, jump, etc.). Then mime with the students that you are sweating and panting. Allow students to have a drink of water if possible. Discuss how you lose water when you exercise, so you need to drink more.

Think!

Why do I sweat?

Explain what sweat is. *What happens when you exercise? My body sweats. Why does your body sweat?* Have students discuss the question in small groups. Check answers as a class. (Possible answers: *My body sweats because it gets too hot. Sweat helps it stay cool.*) Then ask *What do you need when you sweat a lot? Why?* Elicit ideas from the students. (*You need to drink water because sweat has water in it.*)

I Will Know…

Have students do the *I Will Know…* Digital Activity.

What does my body need?

> **Objective:** Learn how sleep and shelter help your body.
>
> **Vocabulary:** *sleep, shelter, remember, learn, grow, safe, warm*
>
> **Digital Resources:** Flash Cards (*healthy, sleep, shelter*), *Lesson 2 Check* (print out 1 per student), *Got it? 60-Second Video*
>
> **Materials:** pictures of animals and their shelters (e.g., dog in a kennel, bird in a nest, bear in a den)

Build Background Display the *healthy* Flash Card and elicit the three things that students learned about their bodies. (Answers: *My body needs food, exercise, and water to stay healthy.*) *What's another thing your body needs?* Elicit ideas. Point to the top right picture. *Today we are going to learn about sleep. Why do we need to sleep?*

Explain

7 **Read. Look at the pictures. Circle the pictures of what your body needs to stay healthy.**

Read the paragraph with the students. Ask them to circle the pictures and check answers as a class. Check understanding. *Why do we need to sleep? How does sleep help our bodies?* (Possible answer: *We grow when we sleep!*)

ELL Content Support

Some living things need sleep to survive. A child between the ages of five and ten years old needs about ten to 11 hours of sleep each night. Sleep allows our bodies and our brains to rest. When we don't get enough sleep, we can feel tired and grumpy, and we can find it hard to think. We can also get sick more easily when we don't get enough sleep.

8 **Read. How does shelter help you? Say with a partner.**

Draw students' attention to the pictures. *Where are the prairie dogs? Where is the boy?* Elicit that they are at home. Read the paragraph and explain that *shelter* means *home*. *Is it safe or dangerous in a shelter?* Safe. *Is it warm or cold in a shelter?* Warm.

Think!

Do our senses help us stay healthy and safe?

Ask the question. Encourage students to think about how we might know not to touch something that is too hot, to

---

7 **Read. Look at the pictures. Circle the pictures of what your body needs to stay healthy.**

Sleep

You also need sleep to stay healthy. **Sleep** helps you remember what you learn during the day. You even grow when you sleep!

8 **Read. How does shelter help you? Say with a partner.**

Shelter

Many living things need shelter. **Shelter** keeps them safe and warm. You need shelter, too. Your house is a shelter. It keeps you safe and warm!

Prairie dogs use holes for shelter.

Your house is a shelter.

Lesson 2 Check · Got it? 60-Second Video · Unit 5 61

wait to cross a street when we see or hear cars, to wash our feet when they are smelly, etc.

Elaborate

Animal Shelters

Show the pictures of the animals one by one. Elicit the names and what students know about them. Then draw students' attention to the animal homes. Say the names of their homes and help students understand that animals need shelter, too. Encourage students to think about why animals need shelter.

Evaluate

Lesson 2 Check   Assessment for Learning

Review the Key Words for Lesson 2 (see Student's Book page 59). Distribute the *Lesson 2 Check* and guide students as they complete it. Check answers as a class. Then ask students to grade their progress on the topic of what their bodies need from 1 to 3: 3 = *I understand what my body needs*; 2 = *I need to study more*; 1 = *I need help!* Encourage students giving themselves a 1 or a 2 to say what they found difficult and what they need to study more.

> **Got it? 60-Second Video**
>
> Play the *Got it? 60-Second Video* to review the unit material.

Let's Investigate!

In this unit, students learn about their senses and what their bodies need. In this lab, they will find out that different areas of our skin are more or less sensitive to pressure.

Let's Investigate! Lab   **How many points can you feel?**

Objective: Students will discover that their skin is more or less sensitive to pressure in some parts of their bodies.

Materials: per small group: a blindfold, a paper clip

Digital Resources: *Let's Investigate!* Digital Lab, *Let's Investigate!* Activity Card (1 per student)

- Explain to students that they are going to find out more about their skin. They will observe that they can feel a lot or less in different parts of their bodies.

- Take a paper clip and unbend it. Shape it into a letter U. (The ends of the paper clip should be about 2 cm apart.)

- Ask a volunteer to wear the blindfold. Press the ends of the paper clip softly against the student's head. Make sure that both ends of the paper clip touch the skin at the same time. *How many points can you feel? One or two?* Elicit the answer and write on the board: *Head: 2.*

- Invite more volunteers to wear the blindfold, and do the same experiment with students' cheeks, arms, and legs. If you wish, give students paper clips and have them try the task in pairs. Write their results on the board. Remind them to be careful to press firmly but gently and to press only on the areas you indicate.

- Explain that when we feel both tips of the paper clip, it means that our skin is very sensitive in that area of our bodies. Have students complete the *Activity Card.*

- Check answers with the students. (Possible answers: *My cheek and my head feel the paper clip a lot. My arm and my leg do not feel the paper clip a lot.*)

Teacher Time-Saving Option: Show the *Let's Investigate!* Digital Lab as an alternative to the hands-on lab activity.

Unlock the Big Question

Have students refer to the Big Question on the Unit Opener page. In pairs have them recall what they have learned about their senses and their bodies. Have pairs complete questions 6 and 7 on the *Activity Card.*

Let's Investigate!

How many points can you feel?

1. Unbend.
2. Make a letter U.
3. Blindfold.
4. Press.
5. Ask and say.

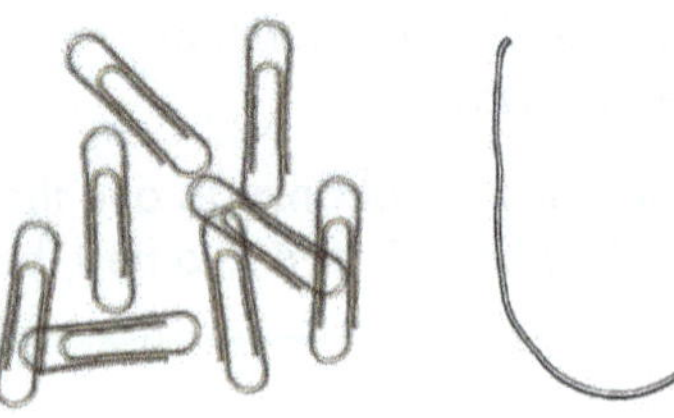

Class Project: Eat a Rainbow

Materials: large sheet of paper, permanent marker, magazine or Internet pictures of different types of fruit, vegetables, rice, pasta, bread, etc. (in color), glue

Cut out the pictures. Draw a big rainbow on the sheet of paper. Make sure that the bands of the rainbow are wide enough for students to stick pictures in them.

Write *Healthy Food* on the board and brainstorm a few ideas. Then show the pictures you have brought and elicit or explain what they are. Put the rainbow on the floor or on the board. Choose a picture of a purple food (e.g., grapes) and invite a student to glue it on the outer band of the rainbow. Continue until the rainbow is full of pictures: red, orange, yellow, green, blue, purple. Encourage students to eat a "rainbow" every day.

Unit 5 Review

What am I like?

Digital Resources: Print out 1 of each per student: *Got it? Self Assessment, Got it? Quiz*

Evaluate

Strategies for Targeted Review

The following are strategies for providing targeted review for students if they encounter challenges with the content.

Lesson 1 What are the five senses?

Question 1

If... students are having difficulty marking the pictures, then... direct students to review Lesson 1. You can also use the *see, hear, touch/feel, taste,* and *smell* Flash Cards to review the vocabulary.

Lesson 2 What does my body need?

Question 2

If... students are having difficulty completing the sentences, then... review the key words as a class: *healthy, exercise, grow.* Then ask questions. *What keeps our muscles and bones healthy? What happens when we sweat? What do we do at night that helps us remember what we learn?*

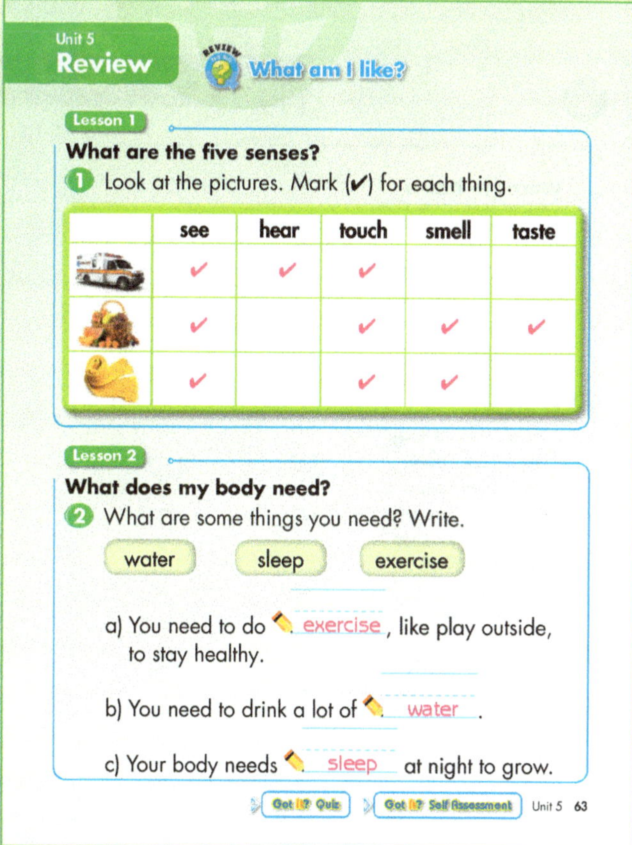

ELL Language Support

Before students start working on the Review activities, have them read each question aloud along with you.

Got it? Self Assessment

Immediately after students have completed the Review activities, distribute a *Got it? Self Assessment* to each student. Have students complete the *Stop! Wait!* and *Go!* statements for each lesson, allowing them to look back through the lesson material if necessary.

Got it? Quiz

Distribute a Unit 5 *Got it? Quiz* to each student. Quizzes may be used for assessing students' understanding of unit concepts as well as for grading purposes.

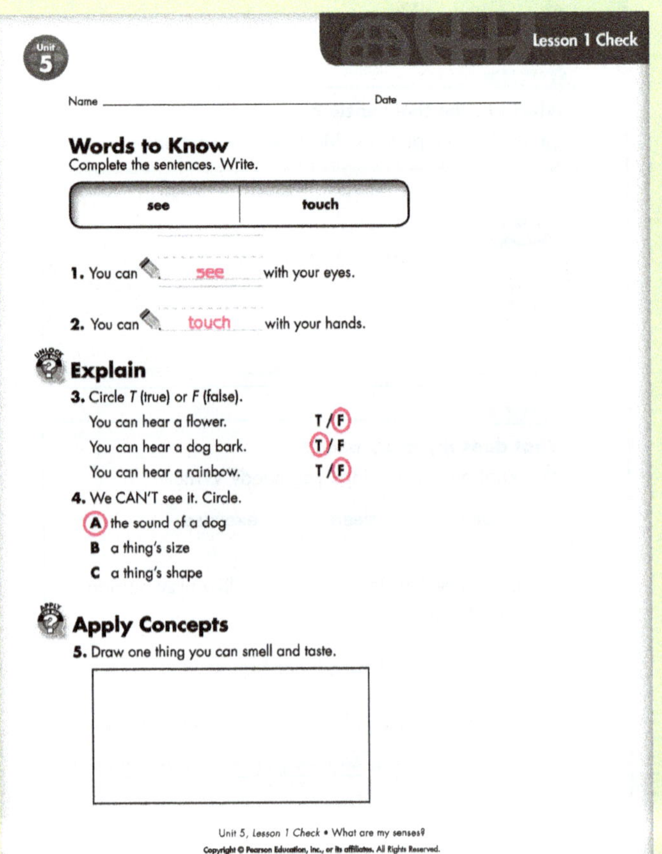

Unit 5

Lesson 1 Check

Name ___________________ Date ___________

Words to Know
Complete the sentences. Write.

| see | touch |

1. You can ✏ _see_ with your eyes.

2. You can ✏ _touch_ with your hands.

Explain

3. Circle T (true) or F (false).
You can hear a flower.   T / (F)
You can hear a dog bark.   (T) / F
You can hear a rainbow.   T / (F)

4. We CAN'T see it. Circle.
(A) the sound of a dog
B a thing's size
C a thing's shape

Apply Concepts

5. Draw one thing you can smell and taste.

Unit 5, Lesson 1 Check • What are my senses?
Copyright © Pearson Education, Inc., or its affiliates. All Rights Reserved.

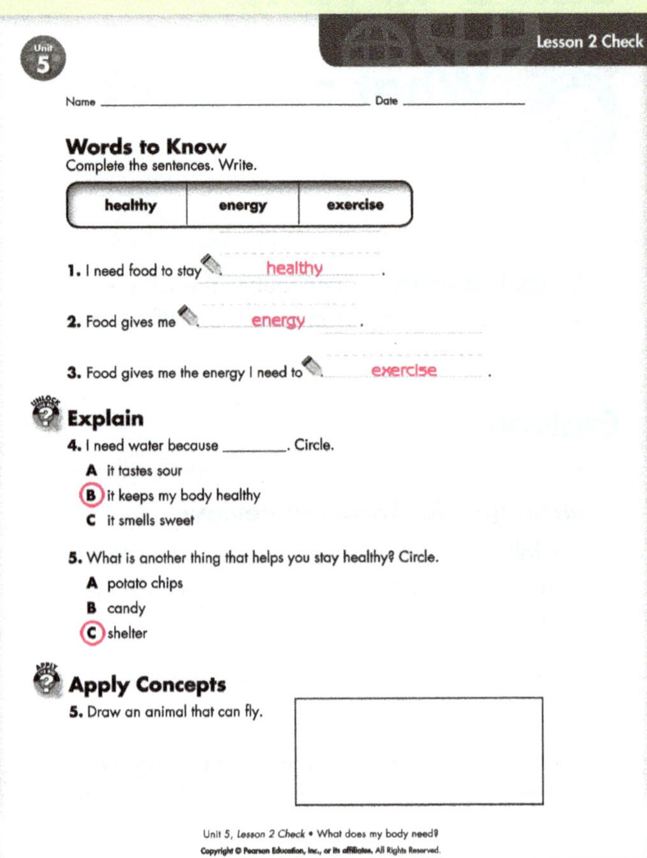

Unit 5

Lesson 2 Check

Name ___________________ Date ___________

Words to Know
Complete the sentences. Write.

| healthy | energy | exercise |

1. I need food to stay ✏ _healthy_ .

2. Food gives me ✏ _energy_ .

3. Food gives me the energy I need to ✏ _exercise_ .

Explain

4. I need water because ________. Circle.
A it tastes sour
(B) it keeps my body healthy
C it smells sweet

5. What is another thing that helps you stay healthy? Circle.
A potato chips
B candy
(C) shelter

Apply Concepts

5. Draw an animal that can fly.

Unit 5, Lesson 2 Check • What does my body need?
Copyright © Pearson Education, Inc., or its affiliates. All Rights Reserved.

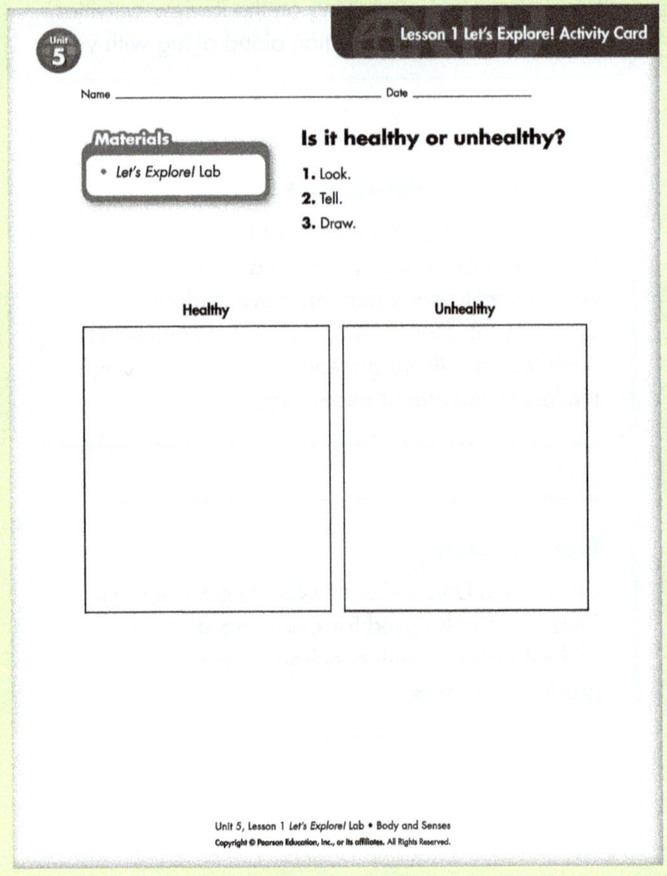

Unit 5

Lesson 1 Let's Explore! Activity Card

Name ___________________ Date ___________

Materials
• Let's Explore! Lab

Is it healthy or unhealthy?

1. Look.
2. Tell.
3. Draw.

Healthy   Unhealthy

Unit 5, Lesson 1 Let's Explore! Lab • Body and Senses
Copyright © Pearson Education, Inc., or its affiliates. All Rights Reserved.

Unit 5

Name _______________________ Date _______________

Analyze and Conclude

6. Circle.

How many points can you feel?		
Body Part	**Number of Points**	
head	1	2
cheek	1	2
arm	1	2
leg	1	2

7. Make a conclusion. Write.

What part of the body feels the paper clip a lot? ✏️ ______

What part of the body does not feel the paper clip a lot? ✏️ ______

Unit 5, *Let's Investigate!* Lab • Body and Senses

Unit 5

Name _______________________ Date _______________

Got it? Self Assessment

Complete the statements for each lesson.

Lesson 1 What are my senses?

⏹ **Stop!** I need help with _________________________
____

⏸ **Wait!** I have a question about _________________

▶ **Go!** Now I know _____________________________
____

Lesson 2 What does my body need?

⏹ **Stop!** I need help with _________________________

⏸ **Wait!** I have a question about _________________

▶ **Go!** Now I know _____________________________
____

Unit 5, *Got it? Self Assessment* • Body and Senses

Unit 5

Name _______________________ Date _______________

Got it? Quiz

1. How many senses do you have? Circle.
   A four
   B five
   C six

2. What body part helps you see?
   A nose
   B eyes
   C ears

3. You use your nose to _________.
   A feel
   B hear
   C smell

4. Our eyes do NOT help us see _________. Circle.
   A the color of something
   B how big something is
   C the smell of something

Unit 5, *Got it? Quiz* • Body and Senses

Unit 5

Name _______________________ Date _______________

5. You can feel if things are _________. Circle.
   A cold
   B rough
   C green

6. What are three things you need to stay healthy?
   A food, sleep, and candy
   B food and energy
   C food, sleep, and water

7. Exercise keeps your _________ strong. Circle.
   A bones and muscles
   B bones and teeth
   C bones and ears

8. Shelter can keep living things _________. Circle.
   A safe
   B smooth
   C young

Unit 5, *Got it? Quiz* • Body and Senses

Unit 5 • Digital Resources and Photocopiables **T63b**

Unit 5 Study Guide

What am I like?

Lesson 1
What are my senses?

- We can use our senses to observe the world.
- We can see, hear, touch, taste, and smell.

Lesson 2
What does my body need?

- My body needs food, exercise, water, sleep, and shelter.
- These things help my body stay healthy.

Review the Big Question

What am I like?

Have students use what they have learned from the unit to answer the question in their own words.

How has your answer to the Big Question changed since the beginning of the unit? What are some things you learned that caused your answer to change?

Make a Concept Map

Draw on the board a concept map like the one shown on this page. With the students, talk through the key ideas from this unit. Invite different students to point to the ideas on the board, miming as possible.

Unit 5 Concept Map

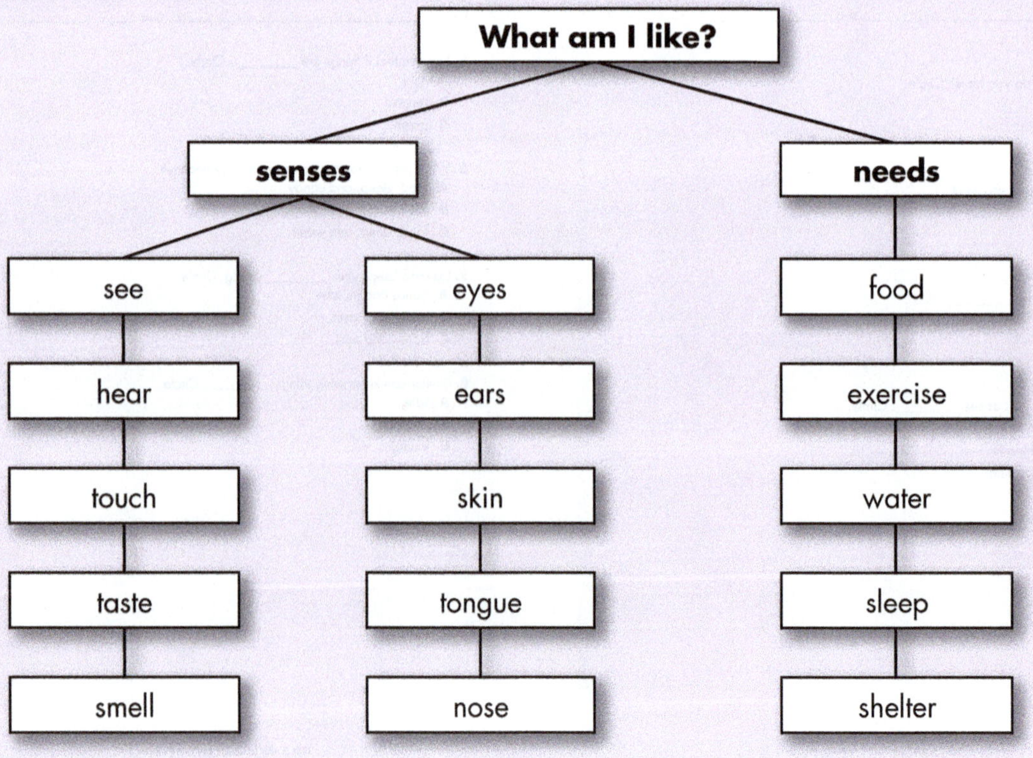

Students can make a concept map to help review the Big Question.

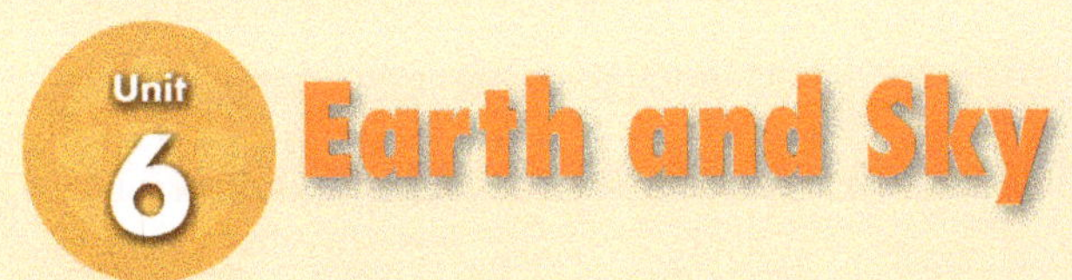

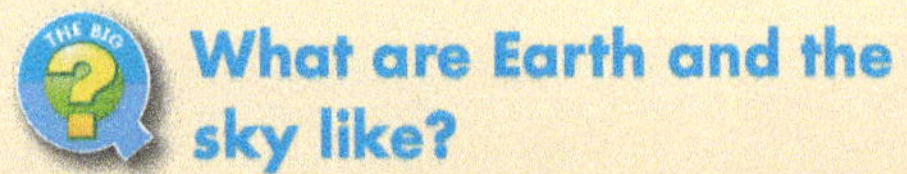
What are Earth and the sky like?

Lesson Plan

Unit Opener & Lesson 1 What makes up Earth?

	Activity	Pages	Time
Engage	• Unit Opener: Think! *Look at the picture. Is it day or night?*	SB p. 64	10 min
	• Unit Opener: Identify bodies of water and landforms.	SB p. 64	10 min
	• Unit Opener: Identify things you can see in the sky.	SB p. 64	10 min
	• Think! *How do you know what is the water and what is the land?*	SB p. 65	10 min
	• Think! *Can some animals live on land and in water?*	SB p. 66	10 min
Explain	• Bodies of water	SB p. 65	30 min
	• Landforms	SB p. 66	30 min
Elaborate	• Ocean Waves	TB p. 65	10 min
Evaluate	• *Lesson 1 Check* (ActiveTeach)	TB p. 75a	10 min
	• Assessment for Learning	TB p. 66	10 min
	• Review (Lesson 1)	SB p. 75	10 min
	• *Got it? Self Assessment* (ActiveTeach)	TB p. 75b	10 min
	• *Got it? Quiz* (ActiveTeach)	TB p. 75b	10 min

Lesson 2 What can you see in the day and night skies?

	Activity	Pages	Time
Engage	• Think! *What else can you see in the day sky?*	SB p. 67	10 min
	• Think! *Is it good or bad to look at the sun?*	TB p. 68	10 min
	• Think! *Why do you like the day sky? Why do you like the night sky?*	TB p. 69	10 min
Explain	• Where the sky is and what you can see in the day sky	SB p. 67	30 min
	• How the sun moves in the sky	SB p. 68	30 min
	• The night sky and what you can see in it	SB p. 69	30 min
Elaborate	• Sun, Earth, and Moon Balloons	TB p. 67	10 min
	• What does the sun really look like?	TB p. 68	10 min
Evaluate	• *Lesson 2 Check* (ActiveTeach)	TB p. 75a	10 min
	• Assessment for Learning	TB p. 69	10 min
	• Review (Lesson 2)	SB p. 75	10 min
	• *Got it? Self Assessment* (ActiveTeach)	TB p. 75b	10 min
	• *Got it? Quiz* (ActiveTeach)	TB p. 75b	10 min

<table>
<tr><th colspan="4">Lesson 3 What is the weather? What are the seasons?</th></tr>
<tr><th></th><th>Activity</th><th>Pages</th><th>Time</th></tr>
<tr>
<td>Engage</td>
<td>• Think! How does the weather change our lives?
• Think! Where does rainwater come from?
• Think! What's your favorite season? Why?
• Think! Does the sky change at different times of the year?</td>
<td>TB p. 70
TB p. 71
TB p. 72
SB p. 73</td>
<td>10 min
10 min
10 min
10 min</td>
</tr>
<tr>
<td>Explore</td>
<td>• Digital Lab: How does weather change? (ActiveTeach)</td>
<td>TB p. 70</td>
<td>30 min</td>
</tr>
<tr>
<td>Explain</td>
<td>• Weather
• Seasons
• Different kinds of seasons
• Got it? 60-Second Video (ActiveTeach)</td>
<td>SB p. 70–71
SB p. 72
SB p. 73
TB p. 73</td>
<td>60 min
30 min
30 min
10 min</td>
</tr>
<tr>
<td>Elaborate</td>
<td>• Different Types of Weather
• Windy, Rainy, Snowy, Sunny
• Four Seasons, Four Trees
• Let's Use a World Map</td>
<td>TB p. 71
TB p. 71
TB p. 72
TB p. 73</td>
<td>10 min
30 min
20 min
10 min</td>
</tr>
<tr>
<td>Evaluate</td>
<td>• Lesson 3 Check (ActiveTeach)
• Assessment for Learning
• Review (Lesson 3)
• Got it? Self Assessment (ActiveTeach)
• Got it? Quiz (ActiveTeach)</td>
<td>TB p. 75a
TB p. 73
SB p. 75
TB p. 75b
TB p. 75b</td>
<td>10 min
10 min
10 min
10 min
10 min</td>
</tr>
<tr>
<td>Lab</td>
<td>• Let's Investigate! What do the day and night skies look like? (ActiveTeach)</td>
<td>SB p. 74</td>
<td>30 min</td>
</tr>
</table>

Flash Cards

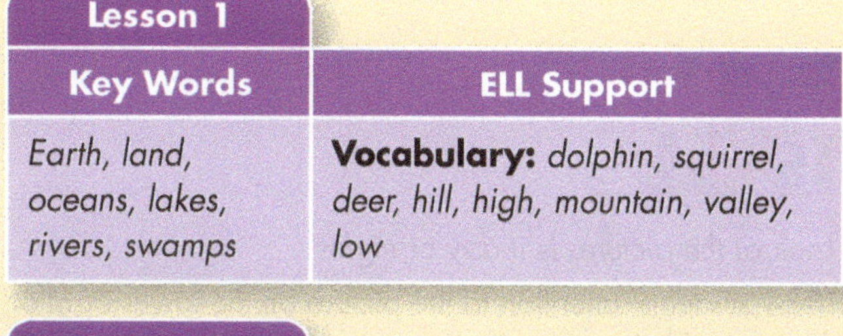

Lesson 1	
Key Words	**ELL Support**
Earth, land, oceans, lakes, rivers, swamps	**Vocabulary:** *dolphin, squirrel, deer, hill, high, mountain, valley, low*

Lesson 2	
Key Words	**ELL Support**
sky, sun, clouds, moon, stars	**Vocabulary:** *light, low, morning, move across, high, noon, evening, night, dark*

Lesson 3	
Key Words	**ELL Support**
weather, sunny, cloudy, clear, rainy, windy, snowy, seasons	**Vocabulary:** *change, What is the weather like?, fall, winter, spring, summer, rainy season, dry season, cold, hot*

Earth and Sky

Unit Objectives

Lesson 1: Students will learn about kinds of land and water on Earth.

Lesson 2: Students will learn what they can see in the day and night skies.

Lesson 3: Students will learn about weather and seasons.

Vocabulary: *Earth, land, oceans, lakes, rivers, swamps, sky, sun, clouds, moon, stars, weather, sunny, cloudy, clear, rainy, windy, snowy, seasons*

Introduce the Big Question

What are Earth and the sky like?

Build Background Write *Home* on the board. *Where is your home?* Elicit or say the students' village, town, or city, then the country, then the continent, and then planet Earth. Provide support as needed. Write *Earth* on the board. Say *Earth is our home.* Then read the title of the unit and the Big Question aloud. Have students focus their attention on the pictures on the page. *Can you see the sky?* Invite students to point to the sky in one of the pictures.

Engage

Think!

Look at the picture. Is it day or night?

Draw students' attention to the picture on the lower right of the page. Elicit or identify the sun and the moon. Then ask *Is it day or night?* Encourage students to share their ideas with one another and to explain their reasoning.

1 **What picture shows water? What picture shows land? Say with a partner.**

Write *Water* and *Land* on the board and check understanding. *Can you drink water?* Yes. *Can you walk on water?* No. *Can you walk on land?* Yes.

Read the questions aloud and have students answer them in pairs. Check answers as a class.

ELL Content Support

Earth is known as the blue planet because it looks blue when seen from space. It looks blue because of the water on its surface. About 71% of Earth's surface is covered by water, and more than 96% of this water is contained within the oceans.

2 **Look at the pictures. Mark (✔) what you can see in the sky.**

Point to the pictures and say the words aloud. *What can you see in the sky?* Have the students mark the correct pictures and check answers as a class.

3 **Is it hot outside? Talk as a class.**

Draw students' attention to the picture. *Is it day or night?* Day. *Can you see the sky?* Yes. *Is it hot?* Elicit answers from students and encourage them to explain their reasoning. Provide vocabulary support as needed. *Is it hot outside today?* Elicit answers from the class.

Think! Again!

Revisit the question *Is it day or night?* Invite students to share their answers and give reasons. Accept all logical answers. (Possible answers: *It's day! The sky is blue, and the sun is in the sky.*)

What makes up Earth?

Objective: Learn about different kinds of water on Earth.

Vocabulary: *Earth, land, oceans, lakes, rivers, swamps, dolphin, squirrel, deer*

Digital Resources: Flash Cards (*Earth, ocean, lake, river*), *I Will Know...* Digital Activity

Materials: Animal Cards (selection of land and water animals, e.g., *camel, coyote, raccoon, tiger, goldfish, octopus, porcupine fish, sea star*)

Unlock the Big Question

Write the following on the board: *I will learn about water. What different kinds of water are there? What animals live in water?*

Build Background Display the *Earth* Flash Card or focus students' attention on the picture of Earth in their books. Have students describe what it looks like.

Explain

1. Read. Is there water close to where you live? Say with a partner.

Read the paragraph aloud. Show the *ocean, lake,* and *river* Flash Cards one by one. Say the words for the students to repeat. *Is there water close to where you live?* Have pairs discuss the question. Elicit answers. If there is an ocean, a lake, or a river nearby, invite volunteers to say whether they go there and any activities they do. *Do you swim there? Do you go to the beach? Do you build sandcastles? Do you ride in a boat on the water?*

2. Look at the pictures. Trace.

Focus students' attention on the pictures. Invite volunteers to say what each picture shows. Then have them trace the words. Monitor and provide support as necessary.

3. Which animals live in water? Say as a class.

Ask students to look at the pictures. Elicit the animals and then ask *Which animals can swim? The dolphin and the fish! Which animals live in water?* Elicit answers from the class. Follow up by asking where students think the deer and squirrel live. *On land!*

Display the Animal Cards and invite students to say whether the animals live in water or not. Keep the cards for the next lesson.

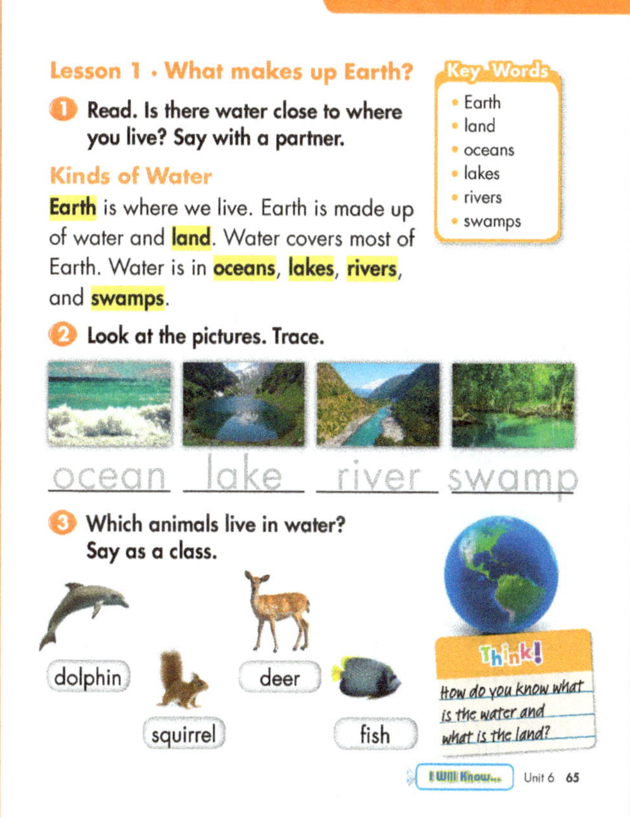

Elaborate

Ocean Waves

Write *wave* on the board and draw one. Check understanding and ask *Where can you usually see waves? In the ocean, in a river, or in a swamp? Ocean.*

Ask a student on one side of the classroom to stand up, raise their arms in the air, and sit down. As one student sits down, the next student stands up and repeats the movement. Have students continue the "wave" around the class.

Think!

How do you know what is the water and what is the land?

Draw students' attention to the picture of Earth. Point to the blue parts. *What color is this? Blue!* Then point to the green parts. *What color is this? Green!* Then read the *Think!* question and allow students to discuss freely. Finally, invite students to say what they think the answers are and why. (Possible answers: *I think the blue part is water. The ocean's water looks blue. I think the land is green. Many plants on land are green.*)

I Will Know...

Have students do the *I Will Know...* Digital Activity.

What makes up Earth?

Objective: Learn about different kinds of land.

Vocabulary: *land, hill, high, mountain, valley, low*

Digital Resources: Flash Cards (*Earth, frog*), Animal Cards (from previous lesson and *duck and duckling, crab, crocodile, sea lion*), *Lesson 1 Check* (print out 1 per student)

Build Background Display the *Earth* Flash Card. Review what students learned about water. Point to a continent. *Is this water? No. What is it?* Elicit *land.*

Explain

4 **Read. What are mountains? Say with a partner. Circle.**

Read the paragraph aloud. Ask students to underline three different kinds of land. (Answers: *hills, mountains, valleys*) Have pairs answer the question. (Answer: *Mountains are land that is very high.*)

ELL Content Support

Around 20% of Earth's land surface is covered by mountains. These form when tectonic plates in Earth's crust move. The highest mountain peak is Everest, which is 8,848 meters high. There are also mountains in the ocean. Many islands are actually the tops of mountains.

5 **What are two kinds of land? Trace.**

Ask the question. Have students trace the words. *Are there hills, mountains, or valleys near here? Have you ever climbed up a mountain?* Encourage volunteers to describe the experience. Prompt students and provide vocabulary support as needed. *Do you see squirrels in the mountains? or deer? Can you see a valley? Are there pine trees?*

6 **What are some animals that live on land? Say as a class.**

Have students identify the animals pictured. *Does a cow live on land? Yes. Does a dog live on land? Yes.* Display the Animal Cards from the previous lesson and invite students to identify the animals that live on land. Have them think of more land animals.

ELL Content Support

Animals like frogs and salamanders are called amphibians. They live part of their lives on land and part in water.

---

4 Read. What are mountains? Say with a partner. Circle.

Kinds of Land

There are many kinds of land on Earth. Hills are land that is high. Mountains are land that is very high. Valleys are land that is low. They have hills or mountains around them.

5 What are two kinds of land? Trace.

    mountain   valley

6 What are some animals that live on land? Say as a class.

Think!

Can some animals live on land and in water?

What are some animals that live in water? Dolphins and fish. What helps fish swim? Their fins. Display the *duck* Animal Card. *Do ducks swim? Yes! Do ducks walk? Yes! Do you think they live on both land and in water? Yes!*

Ask students to think of other animals that live on land and in water. Then display the *crab, crocodile, sea lion,* and *goldfish* Animal Cards. *Which of these animals can't live both on land and in water? The goldfish!*

Finally, point to the picture of the frog. *What is it? A frog. Do you think frogs can live on land and in water? Right! Frogs live in water when they are tadpoles. They have tails! They live on land when they're adults! They have legs!*

Evaluate

Lesson 1 Check   Assessment for Learning

Review the Key Words for Lesson 1 (see Student's Book page 65). Distribute the *Lesson 1 Check* and guide students as they complete it. Check answers as a class. Then ask students to grade their progress on the topic of different kinds of water and land from 1 to 3: 3 = *I understand kinds of water and land on Earth;* 2 = *I need to study more;* 1 = *I need help!* Encourage students giving themselves a 1 or a 2 to say what they found difficult and what they need to study more.

What can you see in the day and night skies?

Objective: What can you see in the day and night skies?

Vocabulary: *sky, sun, clouds, moon, stars, light*

Digital Resources: Flash Card (*sky/sun/clouds*)

Materials: three balloons (yellow, blue, white), permanent marker

Unlock the Big Question

Write the following on the board: *I will learn about what I can see in the day and night skies.*

Build Background Ask the students to look outside. *What can you see?* When students mention the sky, write *sky* on the board. *Do you like looking at the sky? Why?* Invite students to share their opinions.

Explain

1 Read. Where is the sky?

Read the paragraph aloud for the students. *Where is the sky?* Allow students to mime as needed. Ask further questions. *Is the sky beautiful? Is it big or small? Can you see the sky from your bedroom?*

2 Look at the pictures. Trace.

Draw students' attention to the three pictures. Elicit what they can see in each one. Read the paragraph aloud and check comprehension. *What can you see in the sky in the day? The sun, clouds, and sometimes the moon.* Display the *sky/sun/clouds* Flash Card and ask *Can you see the sun? Yes. The clouds? Yes. The moon? No.*

3 What does the sun give Earth? Say with a partner.

Read the question aloud and have pairs discuss. Elicit ideas from the class. (Possible answers: *The sun gives light and heat to Earth. It helps plants grow, and it helps us stay healthy.*) *Is the sun always safe?* Invite students to discuss the possible dangers of the sun. For instance, if we stay in the sun too long in summer, our skin can get burned.

4 What can you see in the day sky? Say with a partner.

Read the question aloud and have students think about it in pairs. Elicit ideas from the class. (Possible answer: *We can sometimes see clouds and the moon*

Lesson 2 · What can you see in the day and night skies?

1 Read. Where is the sky?

The Sky

We see the **sky** when we look up outside. You can see different things in the sky at different times.

2 Look at the pictures. Trace.

The Day Sky

The sky is light in the day. The sun gives Earth light. You can see the **sun** in the sky in the day. Sometimes you can see **clouds**. Sometimes you can see the **moon**, too!

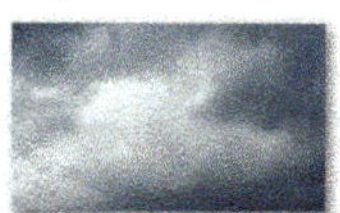
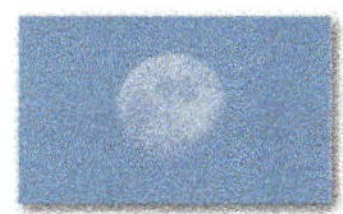

3 What does the sun give Earth? Say with a partner.

4 What can you see in the day sky? Say with a partner.

Key Words
• sky
• sun
• clouds
• moon
• stars

Unit 6 67

---

in the day sky.) You may wish to turn back to the photo on Student's Book page 64 to emphasize that, sometimes, you can see the moon in the day sky.

Elaborate

Sun, Earth, and Moon Balloons

Blow up the three balloons before class. Use a permanent marker to draw the sun on the yellow balloon, Earth on the blue balloon, and the moon on the white balloon. If possible, the sun balloon should be the biggest one and the moon balloon the smallest one.

Show the balloons to the class and explain what they are. Ask a volunteer to stand at the front of the class and hold the sun balloon. Hold the Earth balloon and walk around the student to show the movement of Earth. Start turning as you walk around the student.

Invite another volunteer to hold the moon balloon as you hold the Earth balloon. Have him or her turn as he or she walks around you.

Show the movement of Earth and the moon around the sun with two new volunteers.

Think!

What else can you see in the day sky?

Focus students' attention on the photo of the bird in the picture on the lower right. Have groups discuss the question and then elicit ideas. (Possible answers: *You can see birds, kites, rainbows, airplanes, etc.*)

What can you see in the day and night skies?

> **Objective:** Learn about the sun.
>
> **Vocabulary:** *sun, low, morning, move across, high, noon, evening, night, dark*
>
> **Digital Resources:** Flash Card (*sky/sun/clouds*), *I Will Know…* Digital Activity
>
> **Materials:** real pictures of the sun

Build Background Display the *sky/sun/clouds* Flash Card. Point to the sky, the sun, and clouds in turn. Elicit the words from the class. *Can you see the moon outside today?* Allow students to look outside and check. Elicit answers. Next, explain *high* and *low* by standing up very tall and holding your hands above your head and then crouching down low. Have students follow your movements and say the words after you.

Explain

5 Read. Label the pictures with a partner.

Read the paragraph aloud. Check comprehension by drawing a line to represent the ground on the board and a few clouds to represent the sky. Then draw the sun low on the horizon and high in the sky. Point to the high sun and ask *Is it high or low? High! What time of day is it? Noon!* Then point to the low sun and ask about the sun's position and the time of day. Elicit or explain that it can be morning or evening. Then say *Pretend this sun shows morning.* Draw another sun low on the horizon on the opposite side of the board. *What time of day does this sun show? Evening!* Next, ask *What time of day is it when it's really dark? Night! Can you see the sun at night? No!*

Finally, focus students' attention on the three pictures and the words. Have them label the pictures in pairs. Check answers as a class.

6 How are morning and evening alike? Talk as a class.

Read the question aloud and elicit ideas from the class. Invite students to read the paragraph in exercise 5 again if necessary. (Answer: *The sun is low in the morning and in the evening.*)

7 How do you know it is noon? Talk with a partner.

Read the question aloud and have students discuss it in pairs. Invite students to read the paragraph in

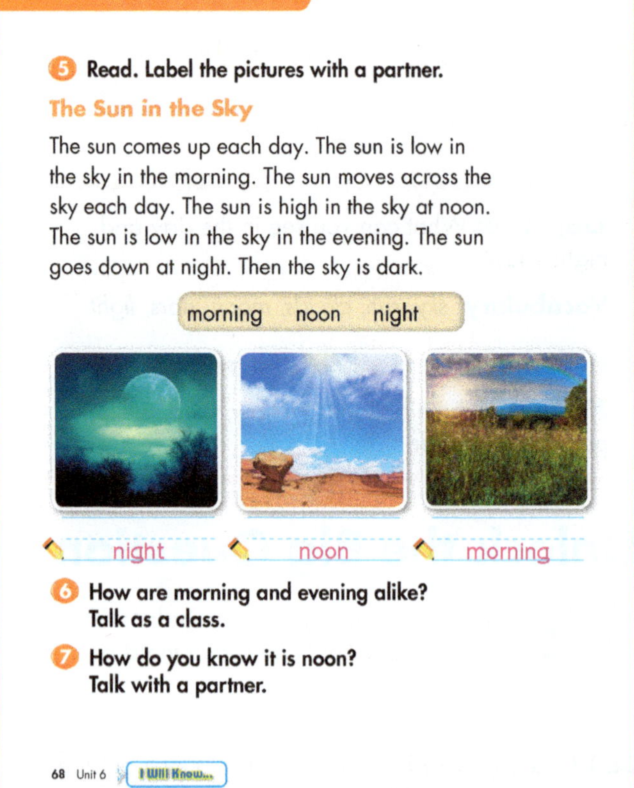

5 Read. Label the pictures with a partner.

The Sun in the Sky

The sun comes up each day. The sun is low in the sky in the morning. The sun moves across the sky each day. The sun is high in the sky at noon. The sun is low in the sky in the evening. The sun goes down at night. Then the sky is dark.

6 How are morning and evening alike? Talk as a class.

7 How do you know it is noon? Talk with a partner.

68   Unit 6   I Will Know…

exercise 5 again if necessary. (Answers: *The sun is high in the sky when it's noon.*)

Think!

Is it good or bad to look at the sun?

Ask the question and have the students discuss it in small groups. Discuss as a class that looking directly at the sun is very dangerous. For example, the sun can burn our eyes the same way it can burn our skin. *How can you protect your eyes from the sun?* Elicit ideas from the class. (Possible answers: *Wear sunglasses and a hat. Don't look at the sun.*)

Elaborate

What does the sun really look like?

Draw a simple drawing of the sun on the board (a circle with rays). *Does the sun look like this?* Elicit ideas from the students. Then show the real photos of the sun. Pass them around for the students to look at. Invite volunteers to describe what they see. Monitor and provide support.

> **I Will Know…**
>
> Have students do the *I Will Know…* Digital Activity.

What can you see in the day and night skies?

> **Objective:** Learn about the night sky.
>
> **Vocabulary:** *stars, sky, night, sun, moon, clouds*
>
> **Digital Resources:** Flash Cards (*sky/sun/clouds, stars*), *Lesson 2 Check* (print out 1 per student)
>
> **Materials:** *optional:* glitter, glue

Build Background Draw a rectangle on the board and say *This is the sky.* Elicit from the students what you can draw in the sky, e.g., the sun, clouds, birds, kites, a rainbow, the moon, etc. Add the items as the students call them out or invite students to draw on the board. *What time of day is it? Is it morning, noon, or evening?* Elicit ideas from the students. Write *The Night Sky* on the board. *How is it different from the day sky?* Invite students to share their ideas.

Explain

8 Read and underline what stars are.

Read the paragraph aloud for students. Check comprehension. *Can you see the sun in the night sky? No. Can you see the (moon) in the night sky? Yes, sometimes. Is it dark at night? Yes.*

9 Look at the pictures. Mark (✔) the picture of the stars.

Focus students' attention on the three pictures. Invite students to describe what they see in each picture. Have them mark the picture of the stars. Check as a class.

ELL Content Support

The light from stars takes a very long time to reach Earth. Some stars that we see in the sky may not actually exist anymore! Stars can have different colors, depending on how hot they are. From low to high temperature, they can be brown, red, orange, yellow, white, or blue. (Brown is the lowest temperature, and blue is the highest temperature.) We think that stars twinkle, but this isn't true. It's the Earth's atmosphere that makes them look like they twinkle.

10 Draw the day sky and the night sky. Share with the class.

Display the *sky/sun/clouds* and the *stars* Flash Cards. *Which is the day sky? Which is the night sky?* Invite students to point to the correct Flash Card.

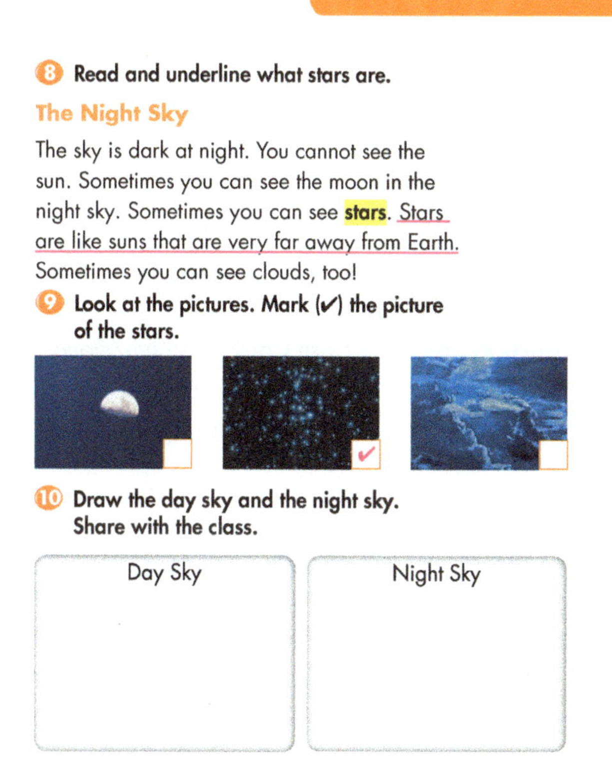

Have the students draw the day sky and the night sky in their books. You may wish to distribute some glitter and glue for the students to add stars in their night sky drawings. Have students show their drawings in small groups.

Think!

Why do you like the day sky? Why do you like the night sky?

Ask the questions and have students discuss them in class. Invite volunteers to say what they like about the sky in the day and at night. Have a class vote on which sky they prefer. Encourage students to say why.

Evaluate

Lesson 2 Check   Assessment for Learning

Review the Key Words for Lesson 2 (see Student's Book page 67). Distribute the *Lesson 2 Check* and guide students as they complete it. Check answers as a class. Then ask students to grade their progress on the topic of the day and night skies from 1 to 3: 3 = *I understand the day and night skies;* 2 = *I need to study more;* 1 = *I need help!* Encourage students giving themselves a 1 or a 2 to say what they found difficult and what they need to study more.

What is the weather?
What are the seasons?

> **Objective:** Learn about the weather.
>
> **Vocabulary:** *weather, sunny, cloudy, clear, rainy, windy, snowy, change*
>
> **Digital Resources:** Flash Cards (*sky/sun/clouds, rainy, snowy, windy*), *Let's Explore!* Digital Lab
>
> **Materials:** different clothing items (e.g., T-shirt, gloves, sunglasses)

Unlock the Big Question

Write the following text on the board: *I will learn about different kinds of weather.*

Build Background Display the Flash Cards on the board. Have the students look at them and say which is most similar to the weather outside. Pre-teach *weather* and discuss whether the weather has changed lately.

Explore

Lets's Explore! Lab How does weather change?

Objective: Learn how the weather changes.

Digital Resources: *Let's Explore!* Digital Lab, *Let's Explore! Activity Card* (1 per student)

- Discuss how the weather can change. Explain that the students are going to observe some ways the weather can change.
- Show the Digital Lab and invite groups to do the activity. Remind students they can refer to information on the board as they go through the activity.
- Alternatively, hand out the *Activity Card* and have students look at the chart. Read the words in the chart aloud and have students mark the words in the first column that describe what the weather is like today. Continue using the chart to record the weather during the next seven to ten days. As a class, discuss whether and how the weather has changed during the period.

Explain

1 Read. Say what weather is with a partner.

Read the paragraph with the students. Point to the Flash Cards and elicit the weather. (Answers: *sunny/ cloudy, rainy, snowy, windy*)

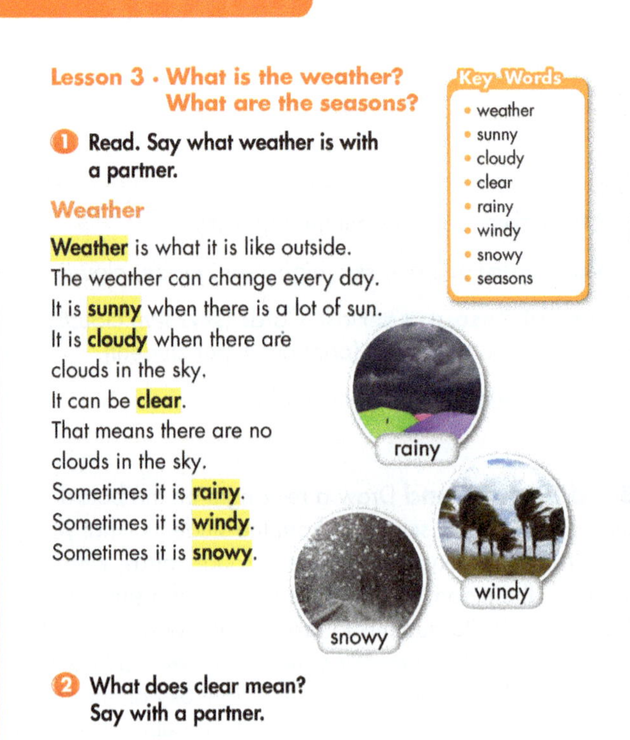

What is weather? Elicit ideas from the students and accept all logical answers. (Possible answer: *What it is like outside.*) Then ask *What is the weather like today?* Have students describe the weather.

2 What does clear mean? Say with a partner.

Point to *clear* in the book and ask *What does* clear *mean?* Have students read the paragraph again and explain the meaning of *clear*. (Possible answer: Clear *means* there are no clouds in the sky) Follow up with some questions about what students can do when it is a clear day. (Possible answers: *I can ride my bike. I can play outside.*)

Think!

How does the weather change our lives?

Point to the Flash Cards on the board. *Is the weather important?* Discuss how the weather affects our lives. Have students talk about what activities they do depending on the weather. Then show the clothing items and elicit in what weather you would wear each one. Have students decide whether the weather is important and how it affects what we do and wear.

What is the weather? What are the seasons?

Objective: Learn more about the weather.

Vocabulary: clear, sunny, snowy, cloudy, What is the weather like?

Digital Resources: Flash Cards (sky/sun/clouds, rainy, windy, snowy), I Will Know… Digital Activity

Build Background Display the Flash Cards on one side of the board. Then draw a happy face and a sad face on the other side. *When it's sunny, are you happy or sad? When it's rainy, are you happy or sad?* Elicit answers from the class. Point out that the weather can make us happy or sad.

Explain

3 Look at the pictures. Match with the words.

Focus students' attention on the pictures. Invite volunteers to read a word aloud. Have them match the pictures with the words. Check answers as a class.

4 What is the weather like? Look and write.

Have the students look at the pictures and describe what they see. Point to each picture and ask *What is the weather like?* Have students label the pictures. Check answers as a class.

5 What is your favorite weather? Why? Talk as a class.

Write the questions on the board. Have two students ask and answer the questions for the whole class to hear. Then have students answer the questions in groups and as a class.

Think!

Where does rainwater come from?

Start by having a class discussion about rain. *Does it rain often where you live? Is it cold or hot when it rains? Do you like rain? Why or why not?* Then ask *Where does rainwater come from?* Have students discuss in small groups.

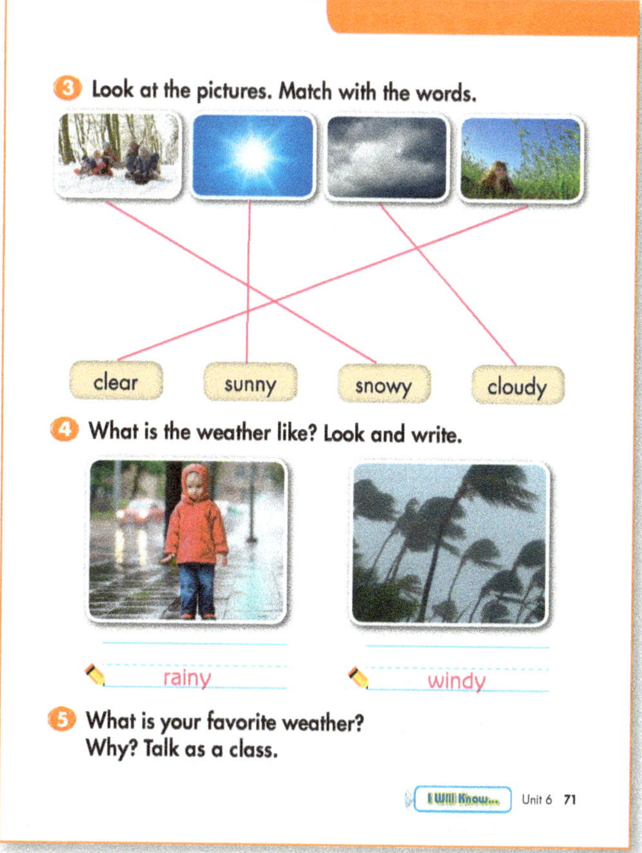

Elaborate

Different Types of Weather

Have students walk around the class and pretend that they are outside in different types of weather. For example, say *It's sunny and hot.* Students can look happy or hot. Say *It's rainy and windy.* Invite students to pretend they're walking in rainy weather, windy weather, snowy weather, and so on.

Windy, Rainy, Snowy, Sunny

Display the Flash Cards at the front of the room. Divide the class into four groups. Teach each group a movement to represent the kind of weather. For windy, students should blow air; for sunny, students should fan themselves; for snowy, they should wrap their arms around them and say *brrr*; for rainy, they should pretend to hold an umbrella. Then collect the Flash Cards and hold up one at a time in random order. The corresponding group stands up and makes its movement. If there is time, have students switch groups and repeat the exercise.

I Will Know…

Have students do the *I Will Know… Digital Activity.*

What is the weather?
What are the seasons?

> **Objective:** Learn about the seasons.
>
> **Vocabulary:** *weather, change, seasons, fall, winter, spring, summer, cold, snowy, trees, plants, flowers, hot*
>
> **Digital Resources:** Flash Cards (*sky/sun/clouds, rainy, snowy, windy*)
>
> **Materials:** construction paper, tissue paper to decorate trees (white, pink, green, orange, and yellow) or decorations for rainy and dry weather (silver sequins for rain and rivers, sand for dry land or riverbeds, etc.), glue, crayons or colored pencils

Build Background Display the Flash Cards on the board and elicit the weather words. Discuss with the students the advantages of each type of weather. For instance, in clear, sunny weather, we can go out and exercise; rainy weather is good for the plants and flowers; when it's snowy, we can play in the snow; if it's windy, we can fly a kite. Point out that there can be good things about each type of weather.

Explain

6 **Read. What is a season? Say with a partner.**

Read the paragraph aloud for students. Check comprehension by asking *How many seasons are there?* Four. *What are they?* Fall, winter, spring, and summer. *What is a season?* Have students discuss in pairs and elicit answers from the class. (Possible answer: *A season is a time of the year with one kind of weather.*) If your area experiences different kinds of seasons, like rainy and dry seasons, you may wish to focus on these instead.

7 **Look at the pictures. Trace the words.**

Draw students' attention to the pictures and elicit the name of the season for each one. Have them trace the words. Have students describe the weather in each season as a class.

Think!

What's your favorite season? Why?

Ask the question and have students discuss it in pairs. Invite volunteers to talk about their favorite season and encourage students to explain why they like it. (Possible answers: *I like summer. It is hot. I swim. I like winter. I play in the snow.*)

---

6 **Read. What is a season? Say with a partner.**

Seasons

The weather can change at different times of the year. These times of the year are called **seasons**. Four seasons are fall, winter, spring, and summer. In some places, leaves change color in fall. Winter is cold and snowy in some places. In spring, trees and plants can get flowers. Summer can be hot in some places.

7 **Look at the pictures. Trace the words.**

72 Unit 6

Elaborate

Four Seasons, Four Trees

Draw a tree trunk and branches on the board. Hand out the construction paper (one piece per student). Ask students to draw the tree trunk and branches. Monitor and provide help as necessary.

Put the students in groups of four, one student for each season. Hand out the tissue paper according to color and season: white for winter, pink for spring, green for summer, orange and yellow for fall. Show the students how to tear small pieces of tissue, crumple them up, and glue them on the tree branches. Monitor and provide help.

Have students label their pictures and invite the groups to stand up and show their four seasons pictures.

Alternatively, you may wish to modify the exercise to focus on other kinds of seasons, like rainy and dry seasons. Have students draw and decorate skyscapes and landscapes accordingly. For example, you may have students draw rivers and skies and decorate them with silver sequins or glitter for rainy season and decorate riverbeds with sand and color bright yellow suns for dry season.

Lesson 3

What is the weather?
What are the seasons?

Objective: Learn about seasons in different places.

Vocabulary: *season, world, places, rainy season*

Digital Resources: Flash Cards (*sky/sun/clouds, rainy, snowy, windy*), *Lesson 3 Check* (print out 1 per student), *Got it? 60-Second Video*

Materials: world map or globe

Build Background Hold up the Flash Cards one by one. For each one ask *What's the weather like? What's the season?* Explain that students are going to learn about weather in different parts of the world.

Explain

6 Read. How many seasons are there where you live? Talk as a class.

Read the paragraph aloud for students. Display the world map or globe and have students find their country. Invite them to talk about the seasons where they live. *What are the seasons? What is the weather like?*

7 What season is it now? What is the weather like? Draw.

Read the two questions aloud and elicit answers from the students. Have them draw a picture for the season and the weather. Invite volunteers to show their pictures to the class and explain what they have drawn.

Elaborate

Let's Use a World Map

Discuss as a class the weather in different places in the world. Show on the map or globe the countries near the equator and explain that the weather is always warm there. Point to India and discuss the rainy season and the dry season. Point out the North and South Poles and talk about the extreme cold weather. Show the Northern Hemisphere and explain that, when it's summer there, it's winter in the Southern Hemisphere. Divide the class into small groups and invite them to discuss what it would be like to live in places with weather and seasons different from where they live now. (Possible answers: *At the North Pole, it is very cold. There is a lot of snow. You need warm clothes.*) Provide vocabulary support as needed.

---

6 Read. How many seasons are there where you live? Talk as a class.

Seasons in Different Places

Seasons are different around the world. Some places have summer when other places have winter. Some places are too hot to have snow. Some places have a rainy season and a season when it doesn't rain a lot.

7 What season is it now? What is the weather like? Draw.

Think!

Does the sky change at different times of the year?

Think!

Does the sky change at different times of the year?

Invite volunteers to discuss the question and describe how they think the sky can change. (Possible answers: *In spring, there are more clouds in the sky. In summer, there is more sun. In rainy season, there is more rain.*) Display the Flash Cards to help students.

Evaluate

Lesson 3 Check   Assessment for Learning

Review the Key Words for Lesson 3 (see Student's Book page 70). Distribute the *Lesson 3 Check* and guide students as they complete it. Check answers as a class. Then ask students to grade their progress on the topic of weather and seasons from 1 to 3: 3 = *I understand different kinds of weather and seasons*; 2 = *I need to study more*; 1 = *I need help!* Encourage students giving themselves a 1 or a 2 to say what they found difficult and what they need to study more.

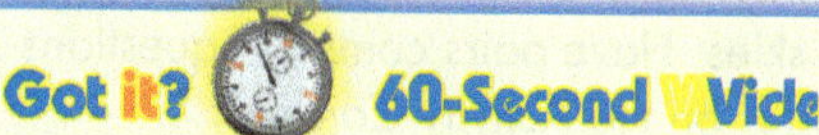

Play the *Got it? 60-Second Video* to review the unit material.

Let's Investigate!

In this unit, students learn about the day and night skies, weather, and seasons. In this lab, they will observe the sky during the day and at night and note the objects they see.

Let's Investigate! Lab — **What do the day and night skies look like?**

Objective: Students will observe the sky and note what objects they see.

Materials: light blue and dark gray construction paper (1 sheet of each per student), crayons, cotton balls, glue

Digital Resources: *Let's Investigate!* Digital Lab, *Let's Investigate! Activity Card* (1 per student)

- Ask *What does the day sky look like? What does the night sky look like?* Elicit answers from the class. Explain that students are going to create a day sky and a night sky.

- Distribute materials and have students create their day sky on the light blue construction paper and their night sky on the dark gray construction paper. Explain that the cotton balls represent the clouds. Students can draw the sun, the stars, and the moon with the crayons. Monitor and provide help as necessary. Have them label their pictures *Day* and *Night*.

- Then invite students to complete the table in their books. Elicit objects (e.g., *clouds, rainbow, moon, stars*) and write them on the board. Have students copy the words in the table and mark (✔) underneath *Day* or *Night* for each object. Discuss which objects we can see both during the day and at night.

- Have students complete the *Activity Card*.

Teacher Time-Saving Option: Show the *Let's Investigate!* Digital Lab as an alternative to the hands-on lab activity.

Unlock the Big Question

Have students refer to the Big Question on the Unit Opener page. In pairs have them recall what they have learned about the day and night skies. Have pairs complete questions 5, 6, and 7 on the *Activity Card*.

Materials
construction paper
crayons
cotton balls and glue

Let's Investigate!

What do the day and night skies look like?

1. Label *Day* and *Night*.
2. Make.
3. Compare.
4. Draw and mark (✔).

What is in the sky?		
Object	**Day**	**Night**
moon	✔	✔
stars		✔
clouds	✔	✔
sun	✔	

74   Unit 6   **Let's Investigate! Lab**

Class Project: Twinkle, Twinkle Little Star

Materials: stars cut out of construction paper, crayons, glitter, glue

Hand out the stars and have students color them. Distribute glitter and glue for the students' stars.

Draw a diamond shape on the board. *Does it look like a star?* Teach the students the popular song (the first verse or both verses) and have them hold up their stars as they sing.

Twinkle, twinkle little star.
How I wonder what you are.
Up above the world so high,
Like a diamond in the sky,
Twinkle, twinkle little star.
How I wonder what you are.

When the blazing sun is gone,
When there's nothing he shines upon,
Then you show your little light.
Twinkle, twinkle through the night.
Twinkle, twinkle little star.
How I wonder what you are.

Unit 6 Review

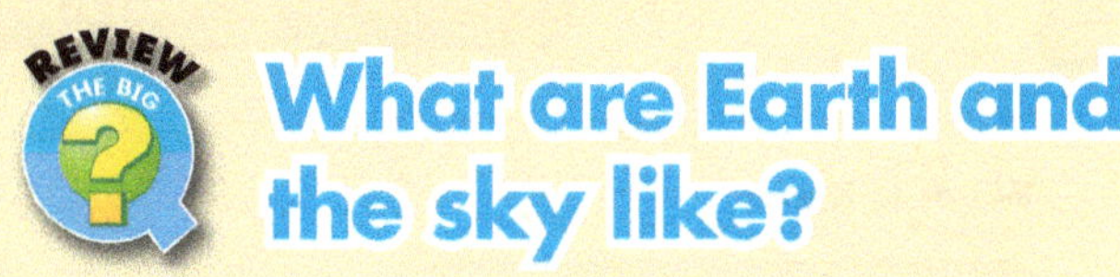

What are Earth and the sky like?

Digital Resources: Print out 1 of each per student: *Got it? Self Assessment, Got it? Quiz*

Evaluate

Strategies for Targeted Review

The following are strategies for providing targeted review for students if they encounter challenges with the content.

Lesson 1 What makes up Earth?

Question 1

If... students are having difficulty labeling the pictures, then... direct students to review Lesson 1.

Lesson 2 What can you see in the day and night skies?

Question 2

If... students are having difficulty identifying the objects they can see in the day and night skies, then... have them refer to the chart they filled in on page 74.

Lesson 3 What is the weather? What are the seasons?

Question 3

If... students are having difficulty completing the sentence, then... direct students to the words and pictures on page 71.

ELL Language Support

Before students start working on the Review activities, have them read each question aloud along with you.

Got it? Self Assessment

Immediately after students have completed the Review activities, distribute a *Got it? Self Assessment* to each student. Have students complete the *Stop! Wait!* and *Go!* statements for each lesson, allowing them to look back through the lesson material if necessary.

Got it? Quiz

Distribute a Unit 6 *Got it? Quiz* to each student. Quizzes may be used for assessing students' understanding of unit concepts as well as for grading purposes.

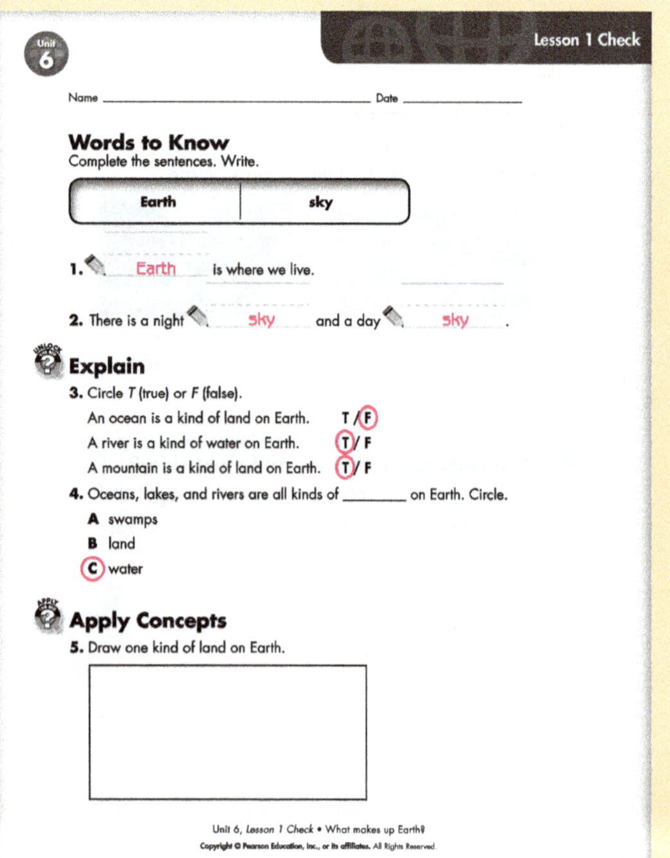

Lesson 1 Check

Name ___________________________ Date ___________

Words to Know
Complete the sentences. Write.

Earth	sky

1. ✎ _Earth_ is where we live.

2. There is a night ✎ _sky_ and a day ✎ _sky_ .

Explain
3. Circle *T* (true) or *F* (false).

An ocean is a kind of land on Earth.   T /(F)

A river is a kind of water on Earth.   (T)/ F

A mountain is a kind of land on Earth.   (T)/ F

4. Oceans, lakes, and rivers are all kinds of _________ on Earth. Circle.

   A swamps

   B land

   (C) water

Apply Concepts
5. Draw one kind of land on Earth.

Unit 6, *Lesson 1 Check* • What makes up Earth?
Copyright © Pearson Education, Inc., or its affiliates. All Rights Reserved.

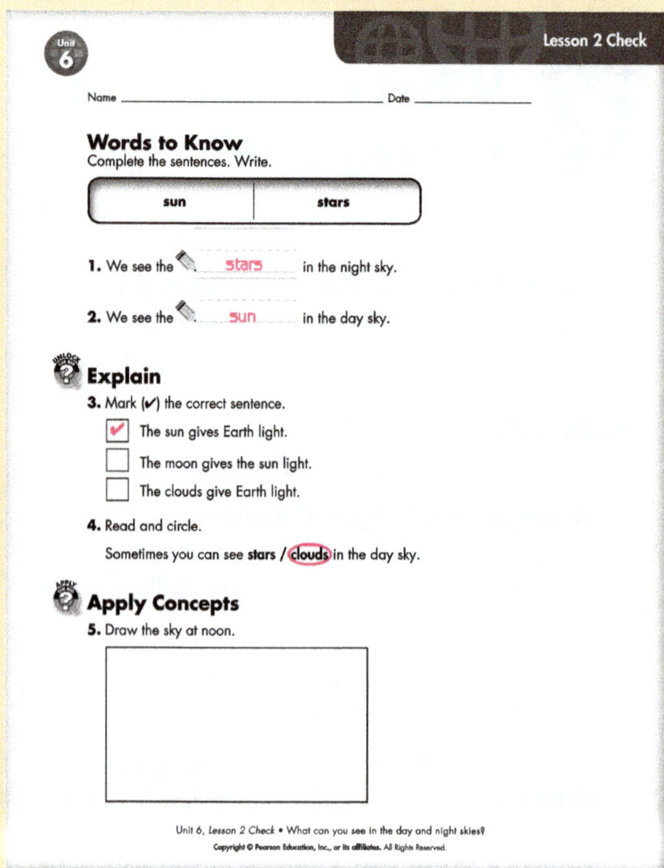

Lesson 2 Check

Name ___________________________ Date ___________

Words to Know
Complete the sentences. Write.

sun	stars

1. We see the ✎ _stars_ in the night sky.

2. We see the ✎ _sun_ in the day sky.

Explain
3. Mark (✔) the correct sentence.

   [✔] The sun gives Earth light.

   [ ] The moon gives the sun light.

   [ ] The clouds give Earth light.

4. Read and circle.

   Sometimes you can see **stars** / (**clouds**) in the day sky.

Apply Concepts
5. Draw the sky at noon.

Unit 6, *Lesson 2 Check* • What can you see in the day and night skies?
Copyright © Pearson Education, Inc., or its affiliates. All Rights Reserved.

Lesson 3 Check

Name ___________________________ Date ___________

Words to Know
Complete the sentences. Write.

weather	season

1. ✎ _Weather_ is what it is like outside. It can change every day.

2. Winter is a ✎ _season_ in some places.

Explain
3. Circle *T* (true) or *F* (false).

When it's sunny, there's a lot of sun.   (T)/ F

When it's windy, there's a lot of wind.   (T)/ F

When it's cloudy, there are no clouds in the sky.   T /(F)

4. What is the order? Number.

   [3] spring

   [4] summer

   [1] fall

   [2] winter

Apply Concepts
5. Draw a cloudy day.

Unit 6, *Lesson 3 Check* • What is weather? What are the seasons?
Copyright © Pearson Education, Inc., or its affiliates. All Rights Reserved.

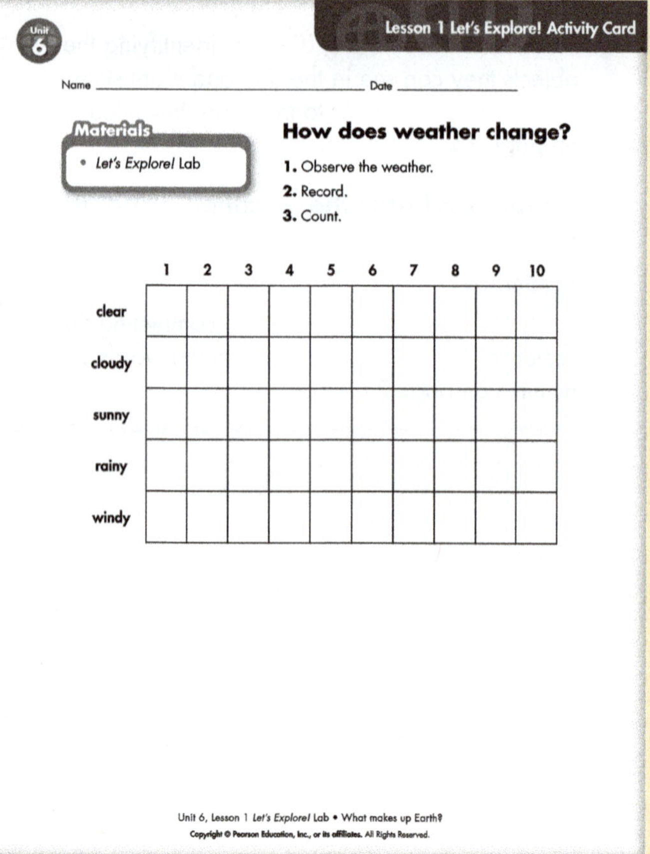

Lesson 1 Let's Explore! Activity Card

Name ___________________________ Date ___________

Materials
- *Let's Explore!* Lab

How does weather change?
1. Observe the weather.
2. Record.
3. Count.

	1	2	3	4	5	6	7	8	9	10
clear										
cloudy										
sunny										
rainy										
windy										

Unit 6, Lesson 1 *Let's Explore!* Lab • What makes up Earth?
Copyright © Pearson Education, Inc., or its affiliates. All Rights Reserved.

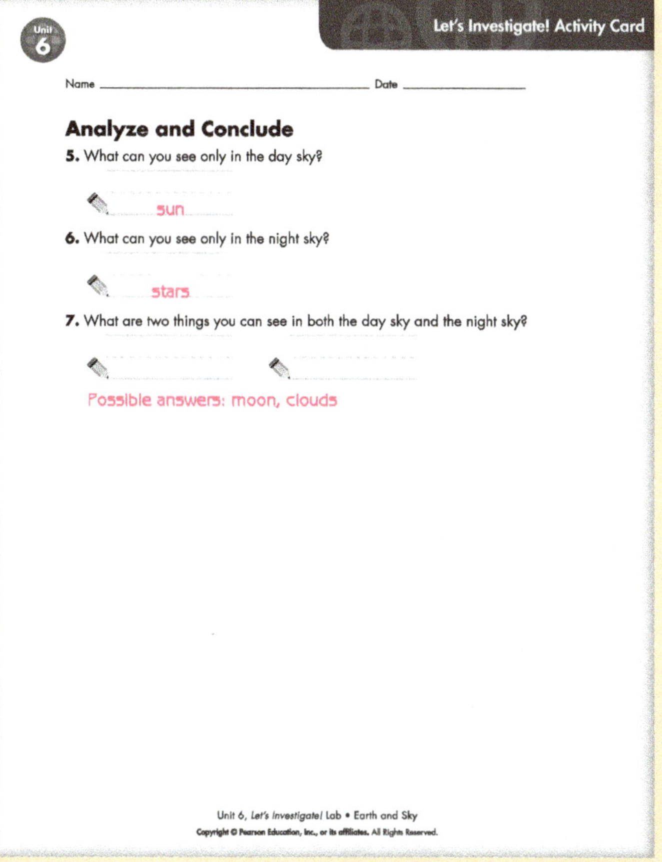

Unit 6

Name ____________________ Date ____________

Analyze and Conclude

5. What can you see only in the day sky?

sun

6. What can you see only in the night sky?

stars

7. What are two things you can see in both the day sky and the night sky?

Possible answers: moon, clouds

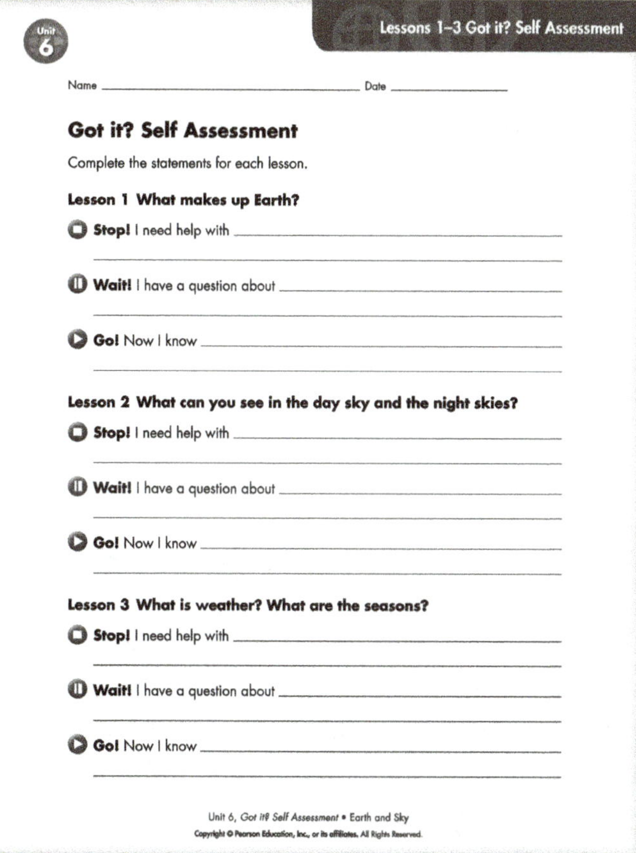

Unit 6

Name ____________________ Date ____________

Got it? Self Assessment

Complete the statements for each lesson.

Lesson 1 What makes up Earth?

Stop! I need help with ____________________

Wait! I have a question about ____________________

Go! Now I know ____________________

Lesson 2 What can you see in the day sky and the night skies?

Stop! I need help with ____________________

Wait! I have a question about ____________________

Go! Now I know ____________________

Lesson 3 What is weather? What are the seasons?

Stop! I need help with ____________________

Wait! I have a question about ____________________

Go! Now I know ____________________

Unit 6

Name ____________________ Date ____________

Got it? Quiz

1. What makes up Earth? Circle.
   A the sky and stars
   (B) land and water
   C stars and the moon

2. What is NOT a kind of water on Earth?
   A a lake
   (B) a valley
   C a river

3. What is NOT a kind of land on Earth?
   (A) a river
   B a valley
   C a hill

4. Which animals live on land?
   (A) dogs
   B dolphins
   C fish

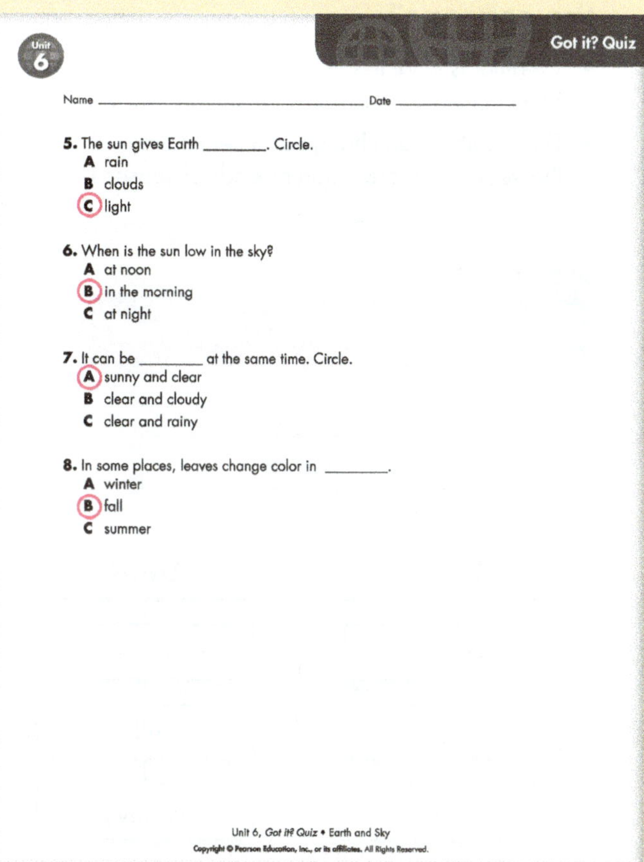

Unit 6

Name ____________________ Date ____________

5. The sun gives Earth ________. Circle.
   A rain
   B clouds
   (C) light

6. When is the sun low in the sky?
   A at noon
   (B) in the morning
   C at night

7. It can be ________ at the same time. Circle.
   (A) sunny and clear
   B clear and cloudy
   C clear and rainy

8. In some places, leaves change color in ________.
   A winter
   (B) fall
   C summer

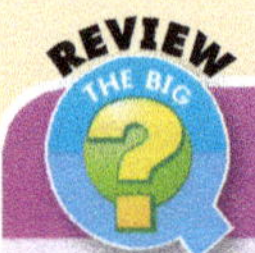

Unit 6 Study Guide

What are Earth and the sky like?

Lesson 1
What makes up Earth?

- Earth is made of water and land.
- Water covers most of Earth. Water is in oceans, lakes, rivers, and swamps.
- Some kinds of land are mountains, hills, and valleys.

Lesson 2
What can you see in the day and night skies?

- The sky is light in the day. We can see the sun and sometimes clouds in the day sky.
- The sky is dark at night. We can usually see the moon and the stars in the night sky.

Lesson 3
What is the weather? What are the seasons?

- Weather is what it is like outside. It can be sunny, cloudy, clear, rainy, windy, and snowy.
- The weather can change at different times of the year. There are different kinds of seasons.

Review the Big Question

What are Earth and the sky like?

Have students use what they have learned from the unit to answer the question in their own words.

How has your answer to the Big Question changed since the beginning of the unit? What are some things you learned that caused your answer to change?

Make a Concept Map

Draw on the board a concept map like the one shown on this page. With the students, talk through the key ideas from this unit. Invite different students to point to the ideas on the board, miming as possible.

Unit 6 Concept Map

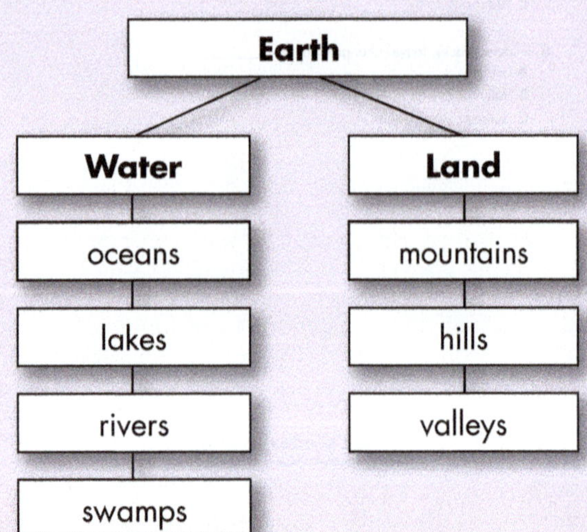

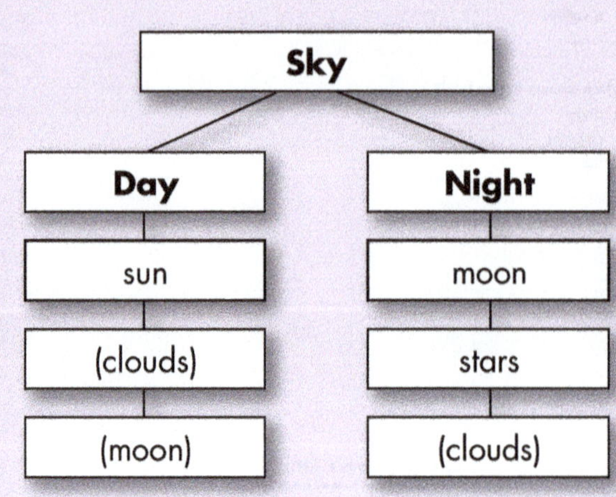

Students can make a concept map to help review the Big Question.

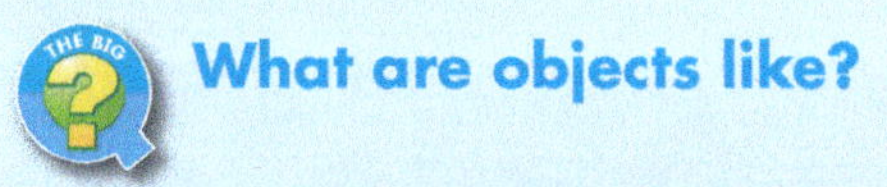

Lesson Plan

Unit Opener & Lesson 1 What are objects made of?		
Activity	**Pages**	**Time**
Engage • Unit Opener: Think! *Can different things have the same colors?*	SB p. 76	10 min
• Unit Opener: Identify the colors and shapes of objects.	SB p. 76	10 min
• Unit Opener: Identify ways to group some objects.	SB p. 76	10 min
• Unit Opener: Identify what some objects are made of.	SB p. 76	10 min
• Think! *What material is a door made of? What material is a handle made of?*	SB p. 78	10 min
• Think! *Where do wood and glass come from?*	TB p. 79	10 min
Explain • Senses and objects	SB p. 77	30 min
• Materials some objects are made of	SB p. 78–79	60 min
Elaborate • Colors and Shapes	TB p. 76	10 min
• Are big objects always heavy?	TB p. 77	10 min
• It's made of…	TB p. 78	10 min
• Where do wood and glass come from?	TB p. 79	10 min
• Flash Lab: Find Objects	SB p. 79	20 min
Evaluate • *Lesson 1 Check* (ActiveTeach)	TB p. 87a	10 min
• Assessment for Learning	TB p. 79	10 min
• Review (Lesson 1)	SB p. 87	10 min
• *Got it? Self Assessment* (ActiveTeach)	TB p. 87b	10 min
• *Got it? Quiz* (ActiveTeach)	TB p. 87b	10 min

Lesson 2 How can you sort objects?		
Activity	**Pages**	**Time**
Engage • Think! *Can you make small groups out of a big group?*	TB p. 81	5 min
Explore • Digital Lab: *How can you sort objects?* (ActiveTeach)	TB p. 80	30 min
Explain • Sorting objects	SB p. 80–81	60 min
• Weighing objects and observing sound	SB p. 82	30 min
Elaborate • Which objects sound similar?	TB p. 80	10 min
• They smell nice. They feel rough. They are heavy.	TB p. 81	10 min
• How can I make different groups?	TB p. 81	10 min
• Make Music with Water	TB p. 82	10 min
• How much does it weigh?	TB p. 82	20 min
Evaluate • *Lesson 2 Check* (ActiveTeach)	TB p. 87a	10 min
• Assessment for Learning	TB p. 82	10 min
• Review (Lesson 2)	SB p. 87	10 min
• *Got it? Self Assessment* (ActiveTeach)	TB p. 87b	10 min
• *Got it? Quiz* (ActiveTeach)	TB p. 87b	10 min

<table>
<tr><td colspan="4" align="center">Lesson 3 How do we use some objects?</td></tr>
<tr><td></td><td align="center">Activity</td><td align="center">Pages</td><td align="center">Time</td></tr>
<tr><td>Engage</td><td>• Think! How many wheels does a car have?
• Think! What can we use square objects for?
• Think! Why are plastic bags bad for our planet?</td><td>TB p. 83
TB p. 83
TB p. 84</td><td>10 min
10 min
10 min</td></tr>
<tr><td>Explain</td><td>• How to use objects based on their properties
• Round objects
• Wood, metal, and glass objects
• Clay and wool as materials
• Got it? 60-Second Video (ActiveTeach)</td><td>SB p. 83
SB p. 83
SB p. 84
SB p. 85
TB p. 85</td><td>15 min
15 min
30 min
30 min
10 min</td></tr>
<tr><td>Elaborate</td><td>• What can we use things with square sides for?
• What is plastic like?
• Go Green: Why are plastic bags bad for our planet?
• What are your clothes made of?</td><td>TB p. 83
TB p. 84
TB p. 84
TB p. 85</td><td>10 min
10 min
10 min
10 min</td></tr>
<tr><td>Evaluate</td><td>• Lesson 3 Check (ActiveTeach)
• Assessment for Learning
• Review (Lesson 3)
• Got it? Self Assessment (ActiveTeach)
• Got it? Quiz (ActiveTeach)</td><td>TB p. 87a
TB p. 85
SB p. 87
TB p. 87b
TB p. 87b</td><td>10 min
10 min
10 min
10 min
10 min</td></tr>
<tr><td>Lab</td><td>• Let's Investigate! Which object is heavier? (ActiveTeach)</td><td>SB p. 86</td><td>30 min</td></tr>
</table>

Flash Cards

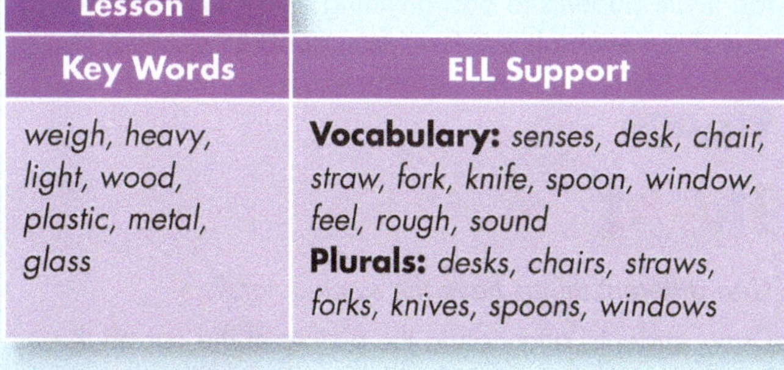

Lesson 1

Key Words	ELL Support
weigh, heavy, light, wood, plastic, metal, glass	**Vocabulary:** senses, desk, chair, straw, fork, knife, spoon, window, feel, rough, sound **Plurals:** desks, chairs, straws, forks, knives, spoons, windows

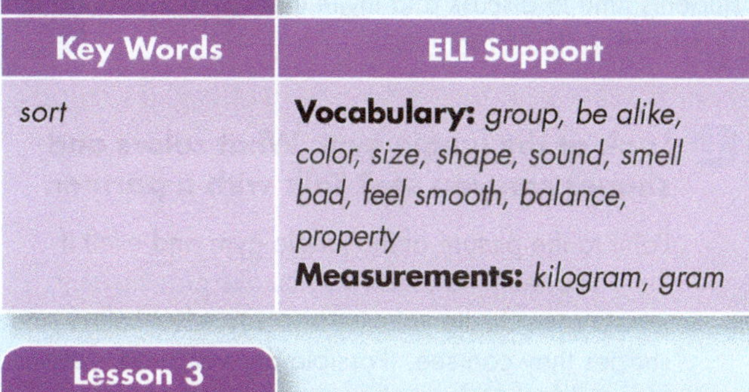

Lesson 2

Key Words	ELL Support
sort	**Vocabulary:** group, be alike, color, size, shape, sound, smell bad, feel smooth, balance, property **Measurements:** kilogram, gram

Lesson 3

Key Words	ELL Support
round, square, strong, see through, clay, sticky, wool	**Vocabulary:** wheel, shape, side, roll, strong, build, furniture, mold, jug, sheep, warm, clothes, sweater

Unit 7 Objects

Unit Objectives

Lesson 1: Students will learn what some objects are like.

Lesson 2: Students will use their senses to sort objects.

Lesson 3: Students will learn some ways to use objects.

Vocabulary: *weigh, heavy, light, wood, plastic, metal, glass, round, square, strong, see through, clay, sticky, wool*

Introduce the Big Question

How can you describe matter?

Build Background Read the title of the lesson and the Big Question aloud. Invite students to choose an object from their pencil cases, backpacks, or the classroom. Have them place the objects on your table. Pick up the objects and elicit what they are and what they are used for. Invite students to add anything else they would like to say about each object. Provide support as needed.

Engage

Think!

Can different things have the same colors?

Draw students' attention to the picture of the car on the lower right of the page. Read the question aloud. Allow students time to discuss and invite them to look around the classroom to find examples.

1 Look at the jungle gym. What colors and shapes can you see? Talk with a partner.

Point to the picture of the jungle gym and elicit if there is one in the area. *Do you like playing on a jungle gym?* Invite volunteers to say what colors and shapes they can see. (Possible answers: *red, yellow, green, blue, pink; circle, square, rectangle*)

2 What are some ways you can group objects? Talk with a partner.

Activate prior knowledge by reminding students that they grouped plants and animals in Unit 3. Look back at the unit as a class. Then have students discuss how they can group objects. Collect different items in the class and have students suggest ways they can group them. (Possible answers: *by color, shape, size*)

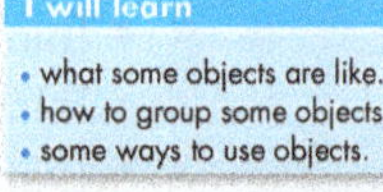

Unit 7 Objects

What are objects like?

I will learn
- what some objects are like.
- how to group some objects.
- some ways to use objects.

1 Look at the jungle gym. What colors and shapes can you see? Talk with a partner.

2 What are some ways you can group objects? Talk with a partner.

3 How many different materials are there in your classroom? Talk as a class.

---

3 How many different materials are there in your classroom? Talk as a class.

Activate prior knowledge by having students remember that materials are what things are made of. *What did the boy use to make his pencil holder?* (Possible answers: *a can, paper, glue, stickers*) *What's a good material for a straw? Plastic!* Elicit or give examples of different materials in the classroom. (Possible answers: *The desks are made of wood. The curtains are made of cotton. The windows are made of glass.*) Invite students to look around the class and discuss the question.

Elaborate

Colors and Shapes

Draw big shapes (circle, triangle, square, rectangle, diamond) on the board. Say the names of the shapes for the students to repeat. Then have them draw the shapes in their notebooks. Invite students to use different crayons to color in the shapes. Then have them label the shapes, e.g., *a yellow triangle.* Write color and shape words on the board.

Think! Again!

Revisit the question *Can different things have the same colors?* Invite students to share their ideas freely. (Answer: *Yes, they can.*) Have students give examples using things they have in their pencil cases or backpacks.

What are objects made of?

Objective: Identify different ways we can use our senses to learn about objects.

Vocabulary: *weigh, heavy, light*

Digital Resources: Flash Cards (*senses, heavy, light*), I Will Know... Digital Activity

Materials: an apple, small heavy objects (e.g., paperweight, dictionary, dumbbell) and big, light objects (e.g., foam cooler, big pillow, sheet of paper, big sponge), a balance

Unlock the Big Question

Write the following on the board: *I will learn more ways to use my senses to learn about objects.*

Build Background Display the *senses* Flash Card. Have students review the senses in pairs, and elicit them from the class: *hear, taste, see, touch, smell.* Hold up the apple. *How can I observe the apple with my senses?* Smell it and elicit *You can smell it.* Look at it and elicit *You can see it.* Touch it and elicit *You can touch it.* Take a bite and elicit *You can taste it. Right! I can use my senses to observe objects and learn about them.*

Explain

1 **Read. How can you tell the size, shape, and color of an object? Say with a partner.**

Read the paragraph. Have pairs discuss, and then check comprehension. *Can you tell the size of something with your ears? No! With your eyes? Yes!*

2 **How can you tell if something is heavy or light? Talk with a partner.**

Show the *heavy* and *light* Flash Cards and elicit which object is heavy (*the box*) and which is light (*the feather*). Display the heavy and light objects on a desk or table. Ask students to decide which are heavy and which are light. At this point, don't let them try to lift the objects.

Ask the question and have students look at the objects in the picture. *Is the feather heavy or light? Light. Is the bus heavy or light? Heavy. How do you know?* Discuss how, sometimes, looking at an object can help us decide if it's heavy or light. Then let students lift the objects you brought. *How can you test your idea? You can feel an object to tell if it is heavy.*

Lesson 1 · What are objects made of?

1 **Read. How can you tell the size, shape, and color of an object? Say with a partner.**

Senses and Objects

You can use your senses to learn about objects. You can hear how objects sound. You can see their size, shape, and color. You can tell how much objects **weigh**. Some objects are **heavy**. Some objects are **light**. You can tell how objects feel. Some objects are rough. Some are smooth.

2 **How can you tell if something is heavy or light? Talk with a partner.**

3 **Look at the pictures. What is rough? Circle. How can you tell?**

Key Words
- weigh
- heavy
- light
- wood
- plastic
- metal
- glass

Finally, if students do not bring it up on their own, ask what the balance does. *It's a tool that tells how much. It weighs things. Right! We can use a balance to tell how heavy or light something is, too. There are some balances big enough to even weigh a bus!*

3 **Look at the pictures. What is rough? Circle. How can you tell?**

Write *rough* and *smooth* on the board. Have students turn to page 56 and look at the pictures in exercise 11. Invite volunteers to say which of the objects is soft (*teddy bear, dog*) and which is rough (*the material*).

Then have students turn to page 77 and look at the two pictures on the lower right. *What is rough?* (*The wood.*) *What is smooth?* (*The rocks.*) *How can you tell?* Elicit or explain that sometimes we can tell just by looking at an object and then check if we are right by touching it.

Elaborate

Are big objects always heavy?

Focus students' attention on the objects on your table. Choose the biggest and smallest objects. Have students lift them and then ask *Is this heavy? Yes. Is it small? Yes. Is this light?* Discuss how sometimes a small object can be heavier than a big object.

That's one reason we feel things or weigh them to find out how heavy they are! Demonstrate with a balance.

What are objects made of?

> **Objective:** Learn what materials some objects are made of.
>
> **Vocabulary:** *wood, plastic, metal, glass, desk, chair, straw, fork, knife, spoon, window*
>
> **Digital Resources:** Flash Cards (*wood, plastic*), *I Will Know…* Digital Activity
>
> **Materials:** objects made of wood (a stick), plastic (a plastic plate), metal (a pot), and glass (a glass), object made of more than one material (eyeglasses)

Build Background Display the objects on a table. Invite students to look at them and say which they think are the heaviest and lightest objects and whether they think they are rough or smooth. Allow students to lift and feel the objects to find out.

Explain

4 **Read. What are some objects made of? Say with a partner.**

Read the paragraph aloud. Check comprehension by pointing to the four photos on the upper right of the page. Elicit or explain what each object is and what it is made of. *What are the logs made of, wood, plastic, metal, or glass?* (Wood!)

Pair students and focus their attention on the objects on your table. Discuss what materials each object is made of. *Which object is made of more than one material?* (The eyeglasses are made of glass and plastic!)

5 **Look at the pictures. Write what each object is made of.**

Have the students look at the pictures and label them with the materials the objects are made of. Monitor and provide help as necessary. Check answers as a class.

ELL Vocabulary Support

You may wish to take the opportunity to review the singular form of the nouns that are in plural in the text. Write *desks, chairs, straws, forks, knives, spoons,* and *windows* on the board. Invite students to say the singular form and focus their attention on the irregular pair *knife-knives*.

---

4 Read. What are some objects made of? Say with a partner.

What Objects Are Made Of

Your senses help you tell what objects are made of. Objects are made of different materials. Some objects are made of **wood**. Desks and chairs can be made of wood. Some objects are made of **plastic**. Straws can be made of plastic. Some objects are made of **metal**. Forks, knives, and spoons can be made of metal. Some objects are made of **glass**. Windows are usually made of glass.

5 Look at the pictures. Write what each object is made of.

metal

glass

plastic

plastic

metal

glass

Think!
What material is a door made of? What material is a handle made of?

Think!

What material is a door made of? What material is a handle made of?

Read the questions aloud and allow students to look at the classroom door closely. Elicit what materials the door and handle are made of. Then have students think about other doors (at home, at school, or in a shopping mall). *Are all doors made of the same materials?* (Possible answer: *No! Some doors are made of glass and metal.*)

Elaborate

It's made of…

Gather students around your table to look at the objects you brought to class again. Invite students to examine each object closely. For example, they might lift the pot and see how heavy it is and try to bend it to see how strong it is, or look through the eyeglasses, and so on. Allow students time to examine several objects and consider some of their properties. Monitor and provide support as necessary. Then have volunteers hold up an item for the whole class and say *It's made of (metal)*. Challenge students to consider what each object can be used for. *You can use (it to cook). You can use (them to see).*

Lesson 1

What are objects made of?

> **Objective:** Learn more about what materials objects are made of.
>
> **Vocabulary:** *wood, glass, metal, fork, desk, window, feel, rough, sound*
>
> **Digital Resources:** Flash Cards (*wood, plastic*), *Lesson 1 Check* (print out 1 per student)
>
> **Materials:** a picture of a tree and a picture of sand

Build Background Show the Flash Cards and elicit the two materials. *What other materials did we talk about? Glass and metal.* Have students find pictures of an object made of wood (e.g., toy horse on page 28), an object made of plastic (e.g., ruler on page 22), an object made of glass (e.g., glass on page 39), and an object made of metal (e.g., paper clips on page 62).

Explain

6 **Look at the pictures. Match the objects with the materials.**

Invite volunteers to read the words aloud. Ask students to point to the desk in the second picture and the window in the third picture. Have them match the objects with the materials. Check answers as a class.

7 **What objects are made of plastic? Say as a class.**

Read the question aloud and have students look at the pictures. Invite volunteers to say which objects are made of plastic. (Answers: *the straw, the ruler*) Have pairs think of two more objects that are made of plastic. (Possible answers: *toys, a jungle gym, a water bottle, plastic bags*)

8 **Circle *T* (true) or *F* (false). Say as a class.**

Invite volunteers to read a sentence aloud. Decide if the sentences are true or false as a class. For the third sentence, knock on a wooden table or door and then knock on the window to help students notice the different sounds.

Elaborate

Where do wood and glass come from?

Have small groups discuss the question. Then show the pictures of the tree and sand. *Do we get wood from the tree or the sand? Tree! What do you think we get from sand? Glass.* Explain that we can make glass by making sand and other things very hot. You may also wish to discuss that metal comes from rocks and plastic is made from chemicals.

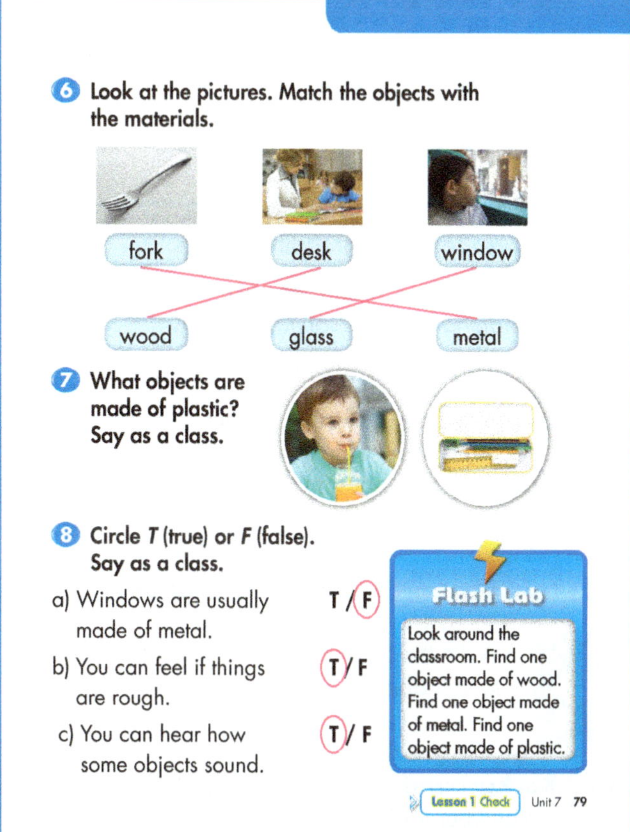

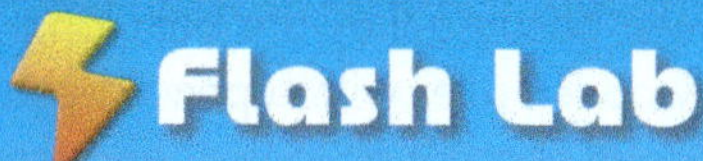

Flash Lab

Find Objects

Put the students in pairs and read the instructions aloud. Have pairs find an object made of wood, and elicit its name from the class. Repeat with an object made of metal and an object made of plastic.

Evaluate

Lesson 1 Check Assessment for Learning

Review the Key Words for Lesson 1 (see Student's Book page 77). Distribute the *Lesson 1 Check* and guide students as they complete it. Check answers as a class. Then ask students to grade their progress on the topic of objects and materials from 1 to 3: 3 = *I understand what objects are made of;* 2 = *I need to study more;* 1 = *I need help!* Encourage students giving themselves a 1 or a 2 to say what they found difficult and what they need to study more.

How can you sort objects?

> **Objective:** Learn about sorting objects.
>
> **Vocabulary:** *sort, group, be alike, color, size, shape, material, weigh, sound, property*
>
> **Digital Resources:** *Let's Explore!* Digital Lab
>
> **Materials:** yellow ball, red modeling clay, feather, metal spoon, white plastic spoon, green, brown, and red crayons, cotton balls, foil, construction paper

Unlock the Big Question

Write the following on the board: *I will learn to sort objects into groups using my senses.*

Build Background Display a few objects from the classroom on your table and have students come close to look at and touch them. Have students discuss how they would pair the objects. For example, they could pair a wooden and a plastic ruler because they are both long. Or, they could pair the wooden ruler with a pencil because they are both made of wood. Accept all logical answers.

Explore

Let's Explore! Lab   How can you sort objects?

Objective: Learn some ways to sort objects.

Digital Resources: *Let's Explore!* Digital Lab, *Let's Explore! Activity Card* (1 per student)

- Activate prior knowledge by eliciting from students what *group* (v) means. *To put things that are alike in a group! Right! Sort is another name for the same thing.*

- Gather students around a desk or table. Display the materials and identify them. Invite students to touch, smell, and lift the items.

- Elicit from volunteers how they would sort or group the items. For example, they may group a ball and a plastic spoon because they are both made of plastic or a feather, a plastic spoon, and cotton balls because they are all light or because they are white.

- Explain to students that *property* means what things are like and that you can sort things based on their properties, e.g., color, size, what they feel like, what they are made of, etc.

- Show the Digital Lab and invite groups to do the activity. Remind students that they can refer to information on the board as they go through the activity.

- Have students work in pairs to complete the *Activity Card.*

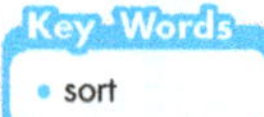

1 **Read. What does sort mean? Say.**

Sort Objects

You can **sort** objects, or put them into groups. You can sort objects by how they are alike. You can sort them by color, size, or shape. You can sort them by the material they are made of or by how much they weigh. You can even sort objects by how they sound!

2 **Look at the pictures. Circle what you can do to sort the objects.**

80   Unit 7   Let's Explore! Lab

Explain

1 **Read. What does sort mean? Say.**

Read the paragraph aloud. Draw students' attention to the highlighted word *sort.* Invite volunteers to explain what it means (*put into groups*). *How can we sort objects?* (Possible answers: *by their color, size, shape, the material they're made of, their weight, how they sound*)

2 **Look at the pictures. Circle what you can do to sort the objects.**

Draw students' attention to the items in the six pictures. Have students identify them and describe their colors, sizes, shapes, weights, materials, etc. Then read the verbs under each picture aloud. Have students circle what they can do to sort the objects. *Can you hear a teddy bear and scarf? No! Can you touch them? Yes!* Check answers as a class.

Elaborate

Which objects sound similar?

Ask students to get a pencil or pen. Have them tap their pencil or pen on different objects in the classroom, for example, a wooden desk, a wooden door, a metal part of a desk or chair, a plastic box, a plastic ruler, a window, etc. Discuss as a class which objects sound similar. Encourage students to discuss why they sound similar. Provide vocabulary support as needed. (Possible answers: *The plastic things sound similar. They are made of the same material. The metal objects sound similar.*)

How can you sort objects?

> **Objective:** Learn more about sorting objects.
>
> **Vocabulary:** *group, smell bad, feel smooth, light*
>
> **Digital Resources:** *I Will Know...* Digital Activity

Build Background Have the students look at the pictures in exercise 3. Give them instructions and ask them to point to a picture. For example, say *Point to something red. Point to something light. Point to something smooth. Point to something that smells nice.* Each time, check which picture the students are pointing to. They might point to different items for the same instruction or the same item for different instructions. Point this out to students.

Explain

3 Look. Make two or more groups. How are the objects in the groups the same?

Put the students in pairs and have them make two or more groups with the objects in the pictures. Encourage them to explain how the objects in a group are the same. Monitor and provide support.

Put pairs in groups of four. Have them discuss the groups they created.

Finally, elicit from the pairs the groups they made and write them on the board. Help students notice how many ways there are to group the objects pictured.

4 Read and draw objects in the three groups.

Read the three headings aloud and give an example for each: *A dirty sock smells bad. An apple feels smooth. A feather is light.* Students can draw their pictures in pairs or on their own. Encourage students to draw on examples from everyday life. Monitor and provide help as necessary.

Elaborate

They smell nice. They feel rough. They are heavy.

Write the three headings on the board. Have students copy them in their notebooks in three columns similar to the columns in exercise 4. Invite students to brainstorm items that smell nice, items that feel rough, and items that are heavy. Then have them draw pictures in their notebooks under each heading. Monitor and provide support.

Have students compare their pictures in small groups.

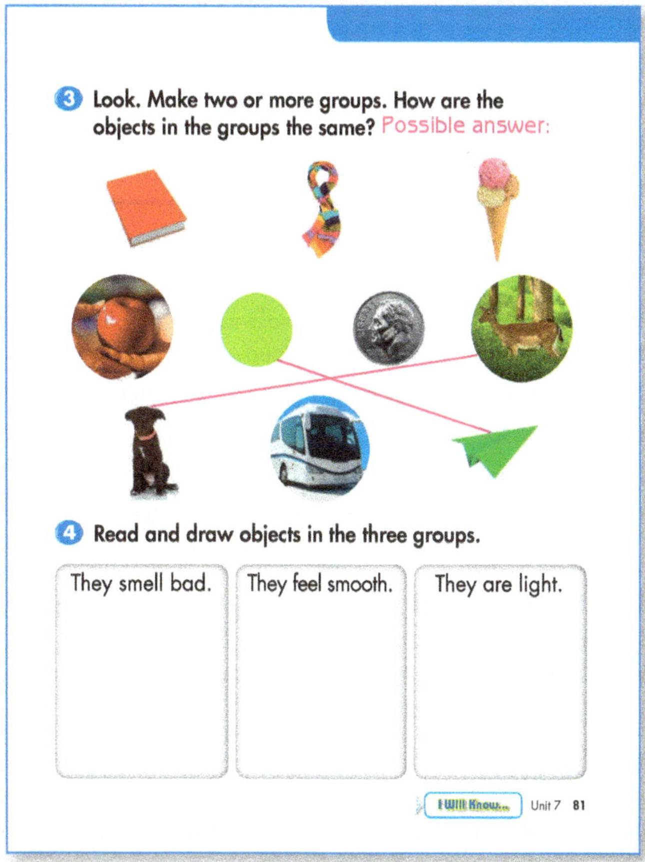

How can I make different groups?

You may wish to challenge students by asking individuals to find additional ways they could sort or group the objects pictured on both pages 80 and 81. Remind students they may look at the list on the board or at the groups in exercise 3 to help them. Or, you may wish to suggest categories, such as food items, cold items, or their favorite items, etc. Invite volunteers to present their ideas to the class.

Think!

Can you make small groups out of a big group? Have students discuss the question fully. If they are having difficulty, ask *Is our class a group? Yes! We are a group of people. Can we make two groups from all the people in the class?* Elicit or explain that two groups could be adults and children. *What are two other groups you can make?* (Possible answer: *Boys and girls.*)

> **I Will Know...**
> Have students do the *I Will Know...* Digital Activity.

How can you sort objects?

Objective: Learn about the weight and sounds of objects.

Vocabulary: *weigh, heavy, balance, sort, light, feather, television, sound, soft, loud*

Digital Resources: Flash Cards (*heavy, light, hear*), *Lesson 2 Check* (print out 1 per student)

Materials: balance and objects that you can weigh, four drinking glasses, a bottle of water

Build Background Ask two volunteers to stand at the front of the classroom and hold up the *heavy* and *light* Flash Cards. Call out items and animals and invite the rest of the class to point to the correct Flash Card.

Explain

5 **Read. Look at the pictures. What do you think weighs a lot? Circle.**

Focus students' attention on the three pictures on the upper part of the page. Elicit or explain what they are: *ice cube, balloon, French horn. What are they made of?* Elicit ideas. (Possible answers: *The ice cube is made of water. The balloon is made of plastic, and it has air inside. The French horn is made of metal.*)

Read the paragraph and the question aloud. Have students circle the item they think weighs a lot. (Answer: *the French horn*) Check answers as a class. Encourage students to explain why they selected the items they did. Accept all logical answers. *How can you test your idea? Right! You can weigh the objects!*

ELL Vocabulary Support

You can take the opportunity to teach *kilogram* and *gram*. If appropriate, invite students to weigh themselves and see how much they weigh.

6 **Read. Look at the pictures. What do you think makes a loud sound?**

Draw students' attention to the three pictures on the lower part of the page. Identify them with the students: *television, airplane,* and *alarm clock.*

Read the paragraph and discuss the question as a class. Discuss which object they think is loudest. *The television can be loud or soft depending on how high the volume is. The alarm clock is usually loud so that we wake up in the morning, but it might not be as loud as an airplane. The airplane is probably the loudest, unless it is very far away.*

5 Read. Look at the pictures. What do you think weighs a lot? Circle.

Weigh Objects

You can weigh things to find out how heavy they are. A balance can help you weigh objects. Then you can sort objects into heavy groups and light groups. A feather can be light. A television can be heavy.

6 Read. Look at the pictures. What do you think makes a loud sound?

Observe Sound

You can sort things by how they sound. Some things make soft sounds. Some things make loud sounds.

Elaborate

Make Music with Water

Fill glasses with different amounts of water and line them up from least to most full. Use a pencil to tap the glasses and have students notice the different sound each glass makes and how the sounds are louder if you tap with more force. *Does it make a loud sound or a soft sound?* Invite volunteers to take turns making some music.

How much does it weigh?

Take out the balance. Make a chart on the board to record the name of each object and its weight. Fill in the chart for one object. Display the others one at a time and have students make predictions about each object's weight compared to the previous one. Invite volunteers to weigh each object and call out its weight. Record the data and discuss the findings as a class. *Were your predictions correct? Are some objects that look heavy light? Are some small objects heavy?*

Evaluate

Lesson 2 Check   Assessment for Learning

Review the Key Words for Lesson 2 (see Student's Book page 80). Distribute the *Lesson 2 Check* and guide students as they complete it. Check answers as a class. Then ask students to grade their progress on the topic of sorting objects using their senses from 1 to 3: 3 = *I understand some ways to sort objects using my senses;* 2 = *I need to study more;* 1 = *I need help!* Encourage students giving themselves a 1 or a 2 to say what they found difficult and what they need to study more.

Lesson 3

How do we use some objects?

Objective: Learn how we use some objects depending on their shape.

Vocabulary: *round, wheel, shape, material, side, roll, square*

Digital Resources: Flash Cards (*round, square*), *I Will Know...* Digital Activity

Materials: objects that are round (e.g., yo-yo, roll of tape, hula hoop, a plate, a globe, a straw hat, etc.), a box

Unlock the Big Question

Write the following on the board: *I will learn how we use some objects.*

Build Background Show the Flash Cards and pre-teach *round* and *square*. Display all the round objects on a table together with the box. *Which object is different?* Have students look at the objects and discuss. *How are all the other objects similar?* Elicit that they are round.

Explain

1 **Read. Why are some round objects good to use as wheels? Say with a partner.**

Pre-teach *wheel* by focusing students' attention on the picture of the bike. Elicit what other objects have wheels, e.g., cars, buses, skateboards, toys, etc.

Read the paragraph and the question. Elicit answers from the class. Use one of the objects you brought in to demonstrate how round objects can roll. *This makes them good to use as wheels!*

2 **Do you think square objects are good to use as wheels? Why? Say as a class.**

Read the question and have students discuss it as a class. Move the box from side to side across a surface to show that it doesn't roll. Discuss why with the students. Elicit *It isn't round!*

3 **What other things can round objects be used for? Say as a class.**

Read the question and focus students' attention on the pictures on the lower part of the page. Elicit or explain what the round objects pictured are. (Answers: *a yo-yo, a hamster wheel, a merry-go-round*)

Lesson 3 · How do we use some objects?

1 **Read. Why are some round objects good to use as wheels? Say with a partner.**

Key Words
- round
- square
- strong
- see through
- clay
- sticky
- wool

How to Use Some Objects

You can use different objects for different things. The shape or material of an object can tell you how you can use it.

Round Objects

Round objects do not have sides. They can roll. You use some round objects as wheels.

2 Do you think square objects are good to use as wheels? Why? Say as a class.

3 What other things can round objects be used for? Say as a class.

I Will Know... Unit 7 **83**

Discuss with students why the round shape of these objects is appropriate in each case. *The yo-yo rolls along the string to help it go up and down. What about the hamster wheel? The hamster wheel rolls, too! Do you think a square would be a good shape for a merry-go-round? No! Right! A square merry-go-round couldn't go around in a circle!*

Think!

How many wheels does a car have?

Draw a car on the board (without the steering wheel). Have pairs discuss the answer. *Why is it good wheels are round? So the car can roll!* Then add the steering wheel to your drawing. Discuss what the steering wheel does and how its shape is appropriate. (*You can turn it in a circle. It turns the front wheels of the car.*)

Elaborate

What can we use things with square sides for?

Hold up the box and ask *Does this box roll? No! Can you put a glass of water on top of a ball? No! Why not? Because it falls off. Can you put a glass on top of the box? Yes! Does the glass fall off the box? No!* Invite students to think of other uses for objects with square sides. At this stage, students do not need to understand what a cube is, only that an object's properties can help us decide how to use it.

I Will Know...

Have students do the *I Will Know...* Digital Activity.

How do we use some objects?

> **Objective:** Learn what wood, metal, and plastic are like and what we can use them for.
>
> **Vocabulary:** *wood, strong, build, furniture, metal, see through, glass*
>
> **Digital Resources:** Flash Cards (*wood, see through*)
>
> **Materials:** object made of wood (e.g., chair), object made of metal (e.g., can), object made of glass (e.g., hand lens), objects made of plastic, some see-through and others not (e.g., bag, cup, spray bottle)

Build Background Show the objects one by one. Invite volunteers to talk about their colors, sizes, shapes, materials, and uses. Discuss which objects are easy to break. *Glass breaks easily, but plastic doesn't.*

Explain

1 **Read. What are some things that are made of wood? Say with a partner.**

Display the *wood* Flash Card and brainstorm objects made of wood. Focus students' attention on the pictures of the table and chair. Read the paragraph aloud and explain *strong*. *Is wood easy to break? No. That's right! It's strong.* Then have students think of other things made of wood and whether they are strong. (Possible answer: *My desk is made of wood. It is strong.*)

2 **Read. What do you think the bus is made of? Why? Talk as a class.**

Read the paragraph and the question aloud. Accept all logical answers. Invite students to think of other objects that are made of metal and are strong. You may wish to point to objects on pages 82 and 83. *Why do we want an airplane to be very strong?* (Possible answers: *A lot of people are in a plane. They are heavy. It needs to be strong.*)

6 **Read. What are some things that are made of glass? Say with a partner.**

Read the paragraph and the question aloud. Pre-teach *see through*. Show the Flash Card and point out the part of the hand lens that is made of glass. *The boy is looking through the glass. He can see through it. Glass is a good material for a hand lens!*

Pair students to brainstorm things made of glass. Invite students to share their lists. Encourage students to say why glass might be a good material for those objects. (Possible answers: *You can see through the windows of the bus. The driver can see other cars.*)

---

4 **Read. What are some things that are made of wood? Say with a partner.**

Wood

Wood can be strong. We use wood to build houses. We use wood to build furniture.

table

chair

5 **Read. What do you think the bus is made of? Why? Talk as a class.**

Metal

Some metal is very strong. We use metal to make things we want to be very strong.

6 **Read. What are some things that are made of glass? Say with a partner.**

Glass

You can see through glass. Glass is a good material for things we want to see through.

84 Unit 7

Elaborate

What is plastic like?

Display the objects made of plastic. Have students touch them. *Is plastic strong? Can you see through it? Can water go through plastic?* Guide students to understand that plastic can be strong and waterproof. *A strong plastic bag is good to carry heavy groceries! Sometimes we can change the shape and color of plastic or even make it see-through, like a water bottle.* Guide students to understand what plastic might be a good for. (Possible answers: *Plastic is good for cups. It doesn't break like glass can.*)

Go Green

Why are plastic bags bad for our planet?

Ask the question and discuss as a class some disadvantages of plastic bags:

- Plastic bags can end up in oceans, lakes, and rivers.
- Some animals, like sea turtles and seabirds, think they are food and swallow them.

Discuss what we can do instead of using plastic bags, for example, use bags you can use again or use a backpack when you go shopping.

How do we use some objects?

> **Objective:** Learn what clay and wool are like and what we can use them for.
>
> **Vocabulary:** *sticky, wool, clay, mold, shape, jug, sheep, warm, clothes, sweater*
>
> **Digital Resources:** Flash Cards (*test, wood, plastic, clay, wool*), *Lesson 3 Check* (print out 1 per student), *Got it? 60-Second Video*
>
> **Materials:** modeling clay, a clothing item made of wool, *optional:* students' own items of clothing

Build Background Display the *test, wood,* and *plastic* Flash Cards. Elicit what materials the objects are made of. (Answers: *metal, wood, plastic*) Invite students to say what they have learned about these materials. If necessary, help them remember by asking questions or allowing them to review pages 77–84 of their Student's Books.

Explain

7 **Read. Why is clay a good material to mold things? Talk as a class.**

Show the modeling clay and allow students to touch it. *What is this material?* Elicit ideas. Then read the paragraph and explain *sticky* and *mold* with the modeling clay. Read the question and discuss with the class. Point to the clay and the jug. Explain that the jug is made of clay. *We can mold things with clay because it's sticky. We can make the shape we want!* Explain that we can make some clay hard by baking it. *It keeps the shape we made. Then we can't mold it into a new shape.*

8 **Read. Why are some clothes made of wool? Talk with a partner.**

Focus students' attention on the three pictures and say the words. Then read the paragraph aloud and check comprehension. *What is the sweater made of? Wool. Where does wool come from? Sheep.*

Have students pass around and touch the clothing item your brought. *Is it rough? Is it soft?* Then have students answer the question. (Possible answer: *Because wool is soft and warm.*) *Do you think wool keeps the sheep warm, too? Yes!*

9 **Do you have any clothes made of wool? Say with a partner.**

Have pairs answer the question. Then have students say when they wear their wool clothes. *Do you wear wool sweaters when it's hot outside? No! Do you wear wool in winter? Yes!* You may wish to invite students to bring a clothing item to class for the next exercise.

7 **Read. Why is clay a good material to mold things? Talk as a class.**

Clay

Clay is sticky. You can mold it. It keeps the shape you make. You can make a jug out of clay.

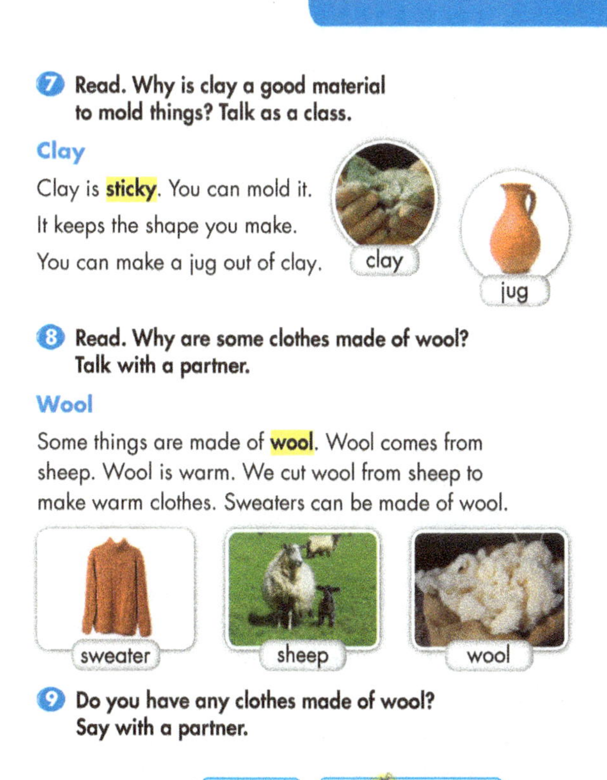

8 **Read. Why are some clothes made of wool? Talk with a partner.**

Wool

Some things are made of wool. Wool comes from sheep. Wool is warm. We cut wool from sheep to make warm clothes. Sweaters can be made of wool.

9 **Do you have any clothes made of wool? Say with a partner.**

Lesson 3 Check | Got it? 60-Second Video | Unit 7 **85**

Elaborate

What are your clothes made of?

Invite students to find out what the clothes they are wearing or brought to class are made of by showing them a clothing label and how to read it. List the materials on the board. Then ask students why they think their clothes are made of those materials. Provide support as necessary. (Possible answers: *My T-shirt is made of cotton. It's light. It's good if it's hot outside. My boots are made of plastic. They are good in the rain. My feet don't get wet.*)

Evaluate

Lesson 3 Check   Assessment for Learning

Review the Key Words for Lesson 3 (see Student's Book page 83). Distribute the *Lesson 3 Check* and guide students as they complete it. Check answers as a class. Then ask students to grade their progress on the topic of how we use some objects based on some of their properties from 1 to 3: 3 = *I understand how we use some objects;* 2 = *I need to study more;* 1 = *I need help!* Encourage students giving themselves a 1 or a 2 to say what they found difficult and what they need to study more.

Play the *Got it? 60-Second Video* to review the unit material.

Let's Investigate!

In this unit, students learn about objects, their materials, and uses. In this lab, they will weigh different objects to find out which one is heavier.

Let's Investigate! Lab   **Which object is heavier?**

Objective: Students will weigh different objects to find out which one is heavier.

Materials: one set per group: balance with plastic cups, crayon, plastic spoon, hand lens, clay, rubber ball, table tennis ball

Digital Resources: *Let's Investigate! Activity Card* (1 per student)

- Show the balance and explain to students how it works. The students will place an object in each plastic cup and find out which object is heavier.

- Model for students with the crayon and the plastic spoon. Have students identify the objects and then invite a volunteer to place them on the balance. Point out that the crayon is heavier than the plastic spoon, and that's why its side of the balance is lower.

- Invite different volunteers to weigh the hand lens and clay and then the rubber ball and table tennis ball. Each time, invite students to say which item is heavier.

- Have students complete the *Activity Card*.

Teacher Time-Saving Option: Show the *Let's Investigate!* Digital Lab as an alternative to the hands-on lab activity.

Unlock the Big Question

Have students refer to the Big Question on the Unit Opener page. In pairs have them recall what they have learned about objects. Have pairs complete question 6 on the *Activity Card*.

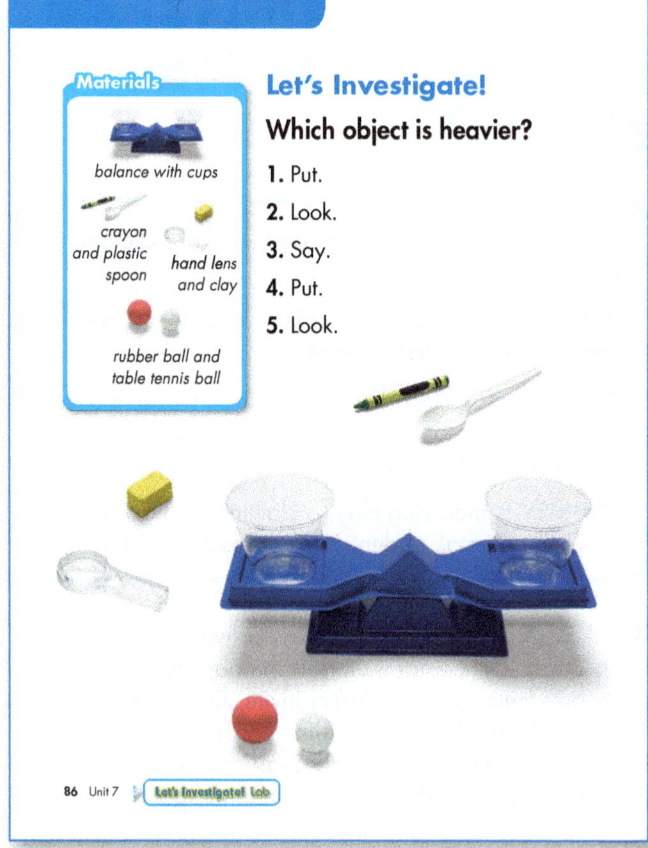

Class Project: Recycling Bins

Materials: four bins or buckets, pictures of objects cut out from magazines, construction paper, glue, sticky tape, garbage (scrap paper, glass bottle, can, plastic bottle)

Display the bins or buckets and explain that they are recycling bins. Explain *recycling* as needed. Tell students that they are going to decorate the bins. Elicit or explain what materials we can recycle: *paper, glass, metal,* and *plastic.* Explain that each bin or bucket is for one of the materials.

Divide the class into four groups (one for each material) and explain that they are going to create small posters to decorate the recycling bins. Distribute the construction paper, the pictures, and the glue. Assign a material to each group and have them stick pictures of objects made of their material to their piece of construction paper. Monitor and provide support as necessary.

Use tape to stick the posters on the bins or buckets. Then show the garbage you have brought and invite students to throw each item into the correct bin. Encourage students to continue to use the bins for their recyclable garbage.

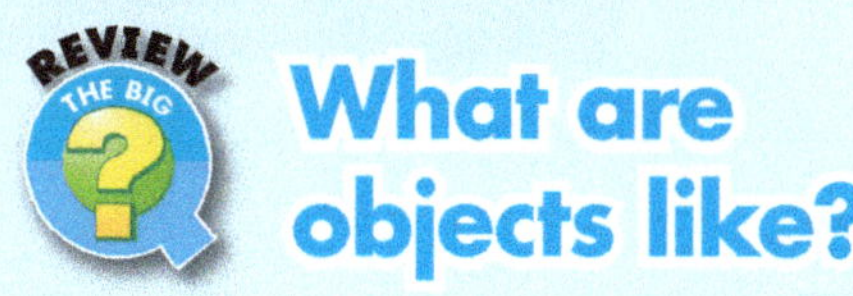

Unit 7 Review

What are objects like?

Digital Resources: Print out 1 of each per student: *Got it? Self Assessment, Got it? Quiz*

Evaluate

Strategies for Targeted Review

The following are strategies for providing targeted review for students if they encounter challenges with the content.

Lesson 1 What is are objects made of?

Question 1

If... students are having difficulty identifying what the object in the picture is made of, then... explain what it is and how it's used. *It's a baseball bat! You hit a ball with it!* Then read the labels aloud and have students write what the bat is made of.

Lesson 2 How can you sort objects?

Question 2

If... students are having difficulty sorting the objects, then... invite volunteers to describe what they look like, or ask questions. *Are the objects round?*

Lesson 3 How do we use some objects?

Question 3

If... students are having difficulty choosing the correct word, then...ask questions to provide clues. *What is wood a good material for? A chair. Is metal a good material for clothes? No! What is wool like? It's warm!*

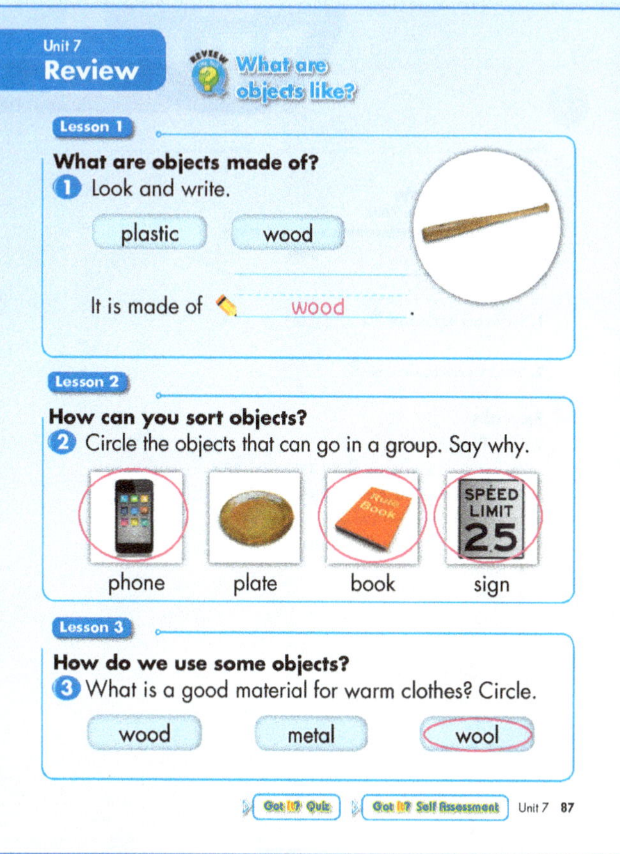

ELL Language Support

Before students start working on the Review activities, have them read each question aloud along with you.

Got it? Self Assessment

Immediately after students have completed the Review activities, distribute a *Got it? Self Assessment* to each student. Have students complete the *Stop! Wait!* and *Go!* statements for each lesson, allowing them to look back through the lesson material if necessary.

Got it? Quiz

Distribute a Unit 7 *Got it? Quiz* to each student. Quizzes may be used for assessing students' understanding of unit concepts as well as for grading purposes.

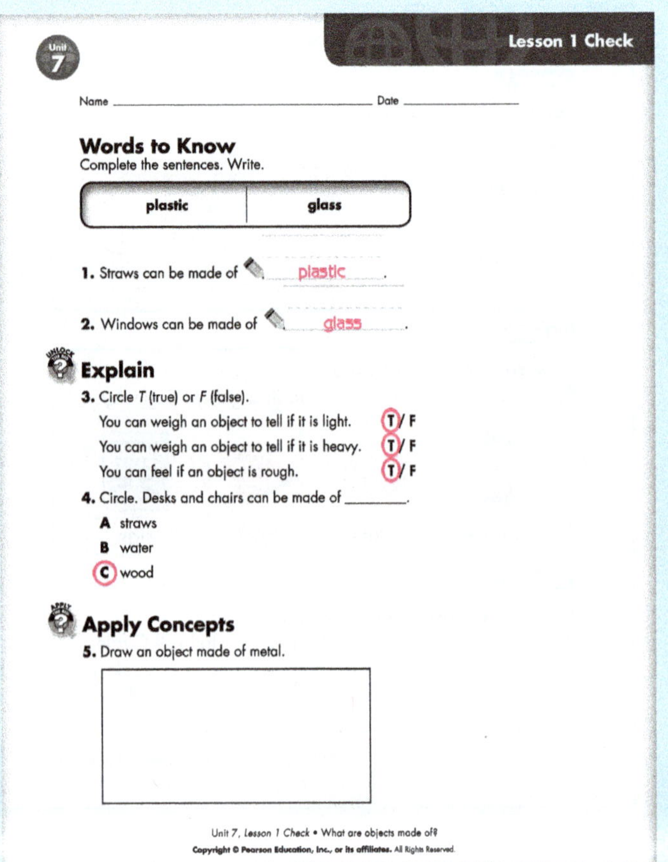

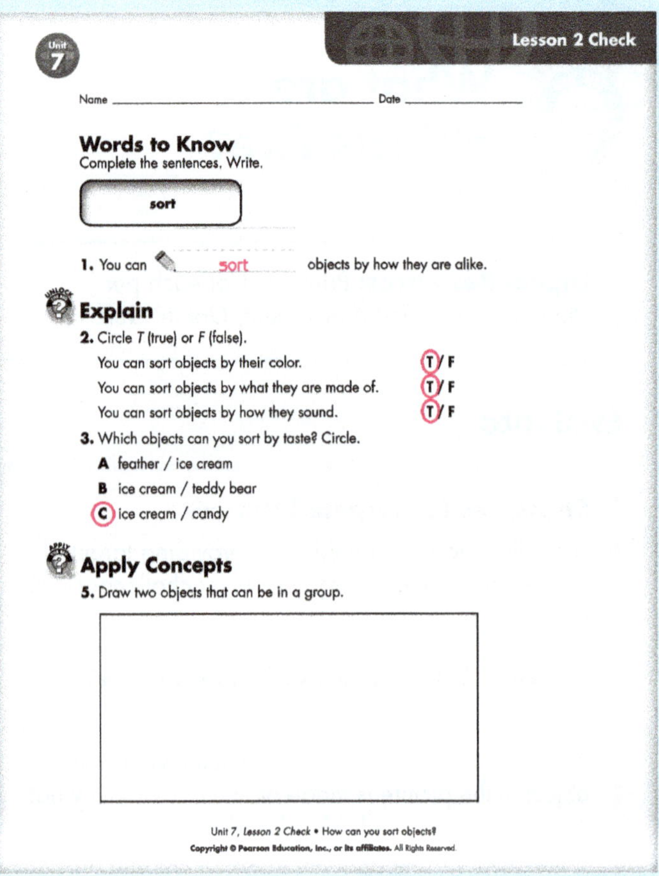

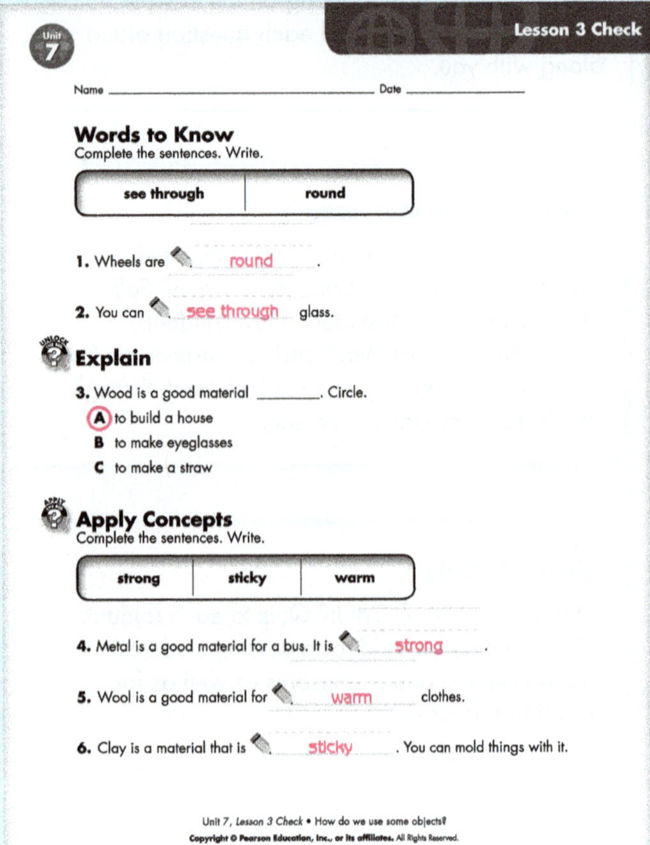

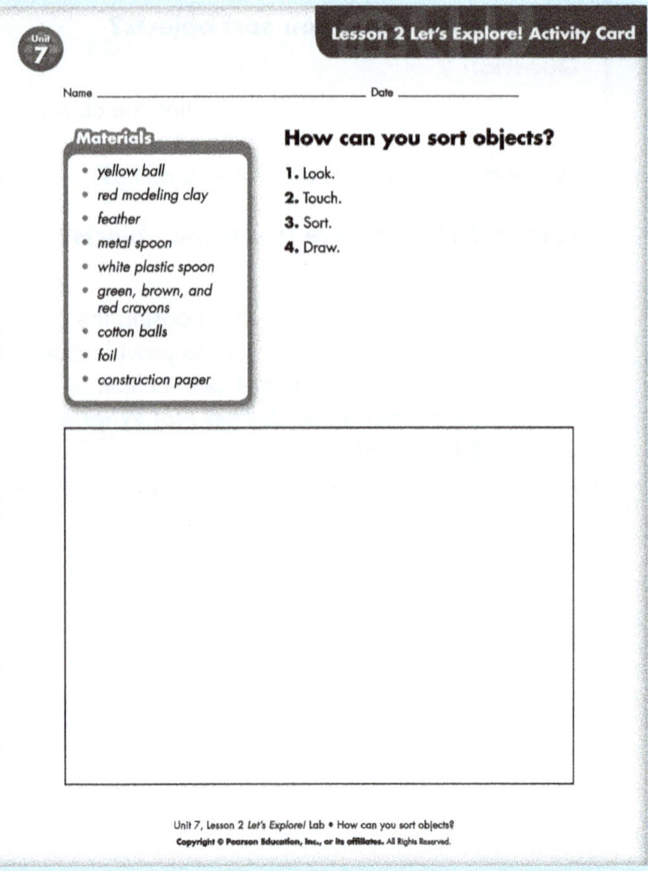

T87a Unit 7 • Digital Resources and Photocopiables

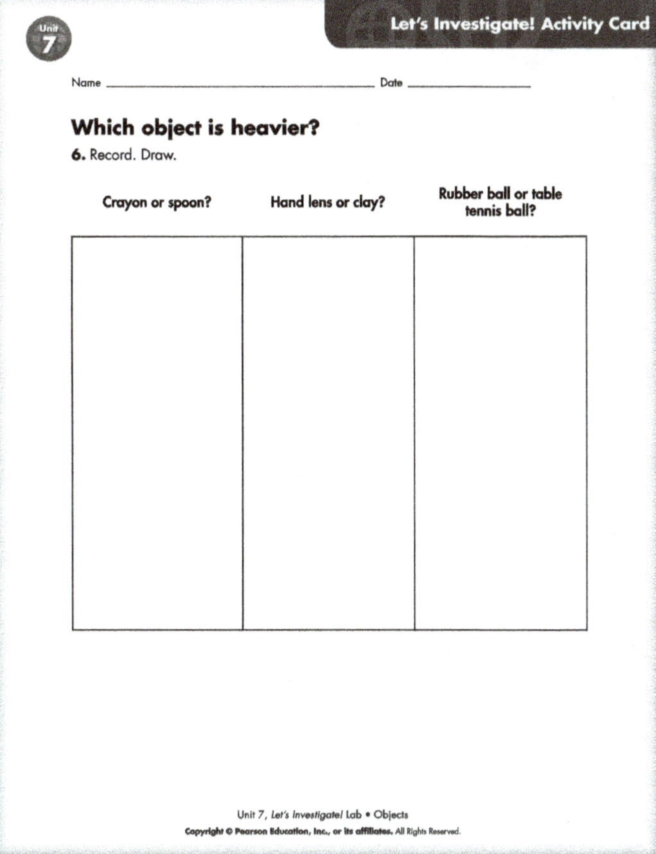

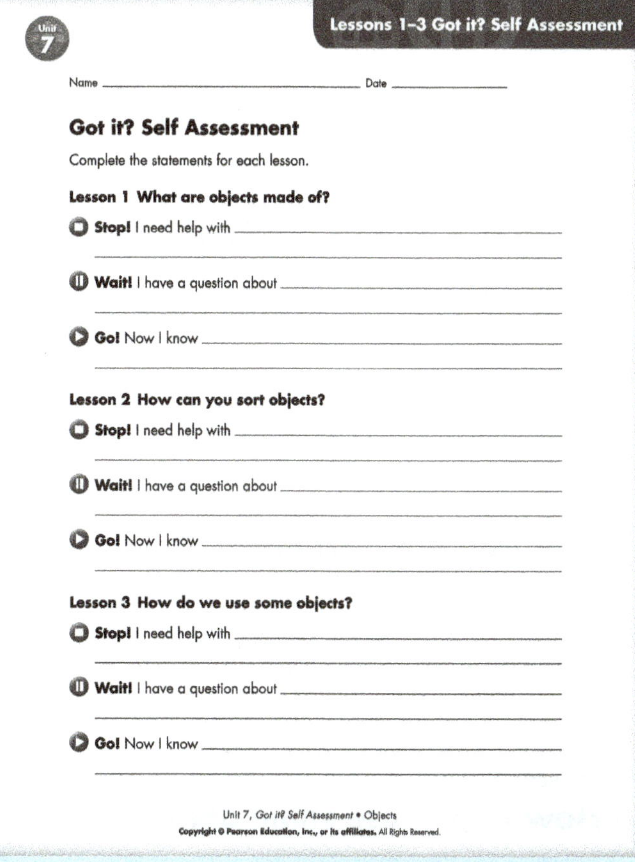

Unit 7

Name _______________ Date _______________

Got it? Quiz

1. Which object is light? Circle.
   A a bus
   B an airplane
   C a feather

2. Which object is rough? Circle.
   A a feather
   B a flower petal
   C a tree trunk

3. What are forks, knives, and spoons usually made of? Circle.
   A metal
   B glass
   C feathers

4. What is your pencil made of? Circle.
   A wool
   B wood
   C glass

Unit 7

Name _______________ Date _______________

5. What is the best way to sort a teddy bear, an ice cream cone, and a wool scarf? Circle.
   A taste
   B touch
   C smell

6. What makes a loud sound? Circle.
   A an ice cube
   B an airplane
   C a feather

7. A round object can be good for _______. Circle.
   A a fork
   B a wheel
   C a building block

8. Wool comes from _______. Circle.
   A horses
   B sheep
   C cows

Unit 7 Study Guide

What are objects like?

Lesson 1
What are objects made of?

- We can say what objects are like. We can say if objects are heavy or light or rough or smooth.
- We can say if objects are made of wood, plastic, metal, or glass.

Lesson 2
How can you sort objects?

- We can sort objects by size, by color, and by shape. We can sort them by the material they are made of, how much weigh, and how they sound.

Lesson 3
How do we use some objects?

- What objects are like can tell us how we can use them.
- We can use some round objects to make wheels. Metal is a good material to make strong things. We can use glass for things we want to see through.

Review the Big Question

What are objects like?

Have students use what they have learned from the unit to answer the question in their own words.

How has your answer to the Big Question changed since the beginning of the unit? What are some things you learned that caused your answer to change?

Make a Concept Map

Draw on the board a concept map like the one shown on this page. With the students, talk through the key ideas from this unit. Invite different students to point to the ideas on the board, miming as possible.

Unit 7 Concept Map

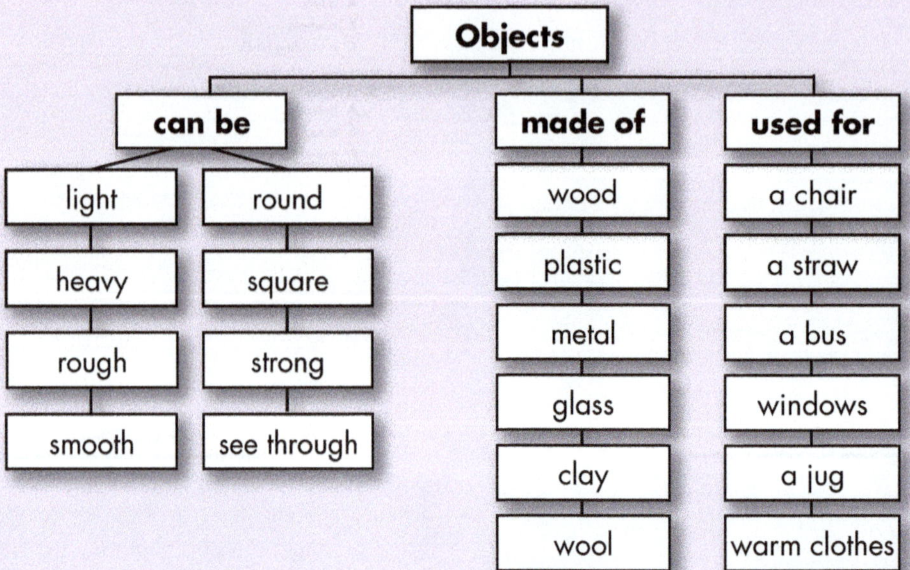

Students can make a concept map to help review the Big Question.

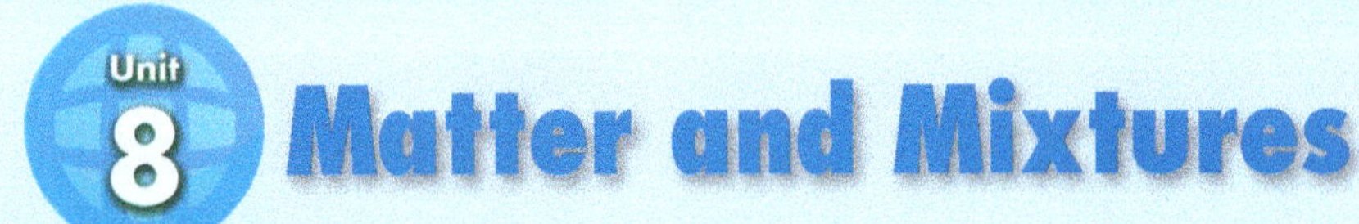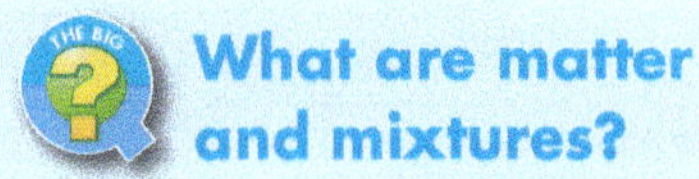

Matter and Mixtures

What are matter and mixtures?

Lesson Plan

Unit Opener & Lesson 1 What are solids, liquids, and gases?		
Activity	**Pages**	**Time**
Engage • Unit Opener: Think! *Is this a mixture?*	SB p. 88	10 min
• Unit Opener: Identify a solid.	SB p. 88	10 min
• Unit Opener: Understand that paper is still paper when cut.	SB p. 88	10 min
• Unit Opener: Identify a mixture.	SB p. 88	10 min
• Think! *Can you make a house out of juice?*	TB p. 89	10 min
• Think! *Is sand a solid?*	TB p. 90	10 min
Explain • Matter	SB p. 89	30 min
• Solids	SB p. 90	30 min
• Liquids	SB p. 91	30 min
• Gases	SB p. 92	30 min
Elaborate • What are clouds made of?	TB p. 88	10 min
• What's the matter?	TB p. 89	20 min
• Building Blocks	TB p. 89	20 min
• Sugar, salt, and flour: are they solids?	TB p. 90	10 min
• Flash Lab: The Shape of Water	SB p. 91	5 min
• How does a straw work?	TB p. 91	10 min
• The Sounds of a Balloon	TB p. 92	10 min
Evaluate • *Lesson 1 Check* (ActiveTeach)	TB p. 99a	10 min
• Assessment for Learning	TB p. 92	10 min
• Review (Lesson 1)	SB p. 99	10 min
• *Got it? Self Assessment* (ActiveTeach)	TB p. 99b	10 min
• *Got it? Quiz* (ActiveTeach)	TB p. 99b	10 min

Lesson 2 How can matter change?		
Activity	**Pages**	**Time**
Engage • Think! *Where can water freeze?*	TB p. 94	10 min
• Think! *Look at the picture. Does the liquid change?*	SB p. 94	10 min
• Think! *Why does the snowman melt?*	SB p. 95	10 min
Explore • Digital Lab: *How do materials change?* (ActiveTeach)	TB p. 80	30 min
Explain • How properties can change	SB p. 93	30 min
• Water can freeze	SB p. 94	30 min
• Water can melt and boil	SB p. 95	30 min
Elaborate • Origami Frog	TB p. 93	20 min
• The Sound of Boiling Water	TB p. 95	10 min
Evaluate • *Lesson 2 Check* (ActiveTeach)	TB p. 99a	10 min
• Assessment for Learning	TB p. 95	10 min
• Review (Lesson 2)	SB p. 99	10 min
• *Got it? Self Assessment* (ActiveTeach)	TB p. 99b	10 min
• *Got it? Quiz* (ActiveTeach)	TB p. 99b	10 min

<table>
<tr><th colspan="4">Lesson 3 What is a mixture?</th></tr>
<tr><td></td><td>Activity</td><td>Pages</td><td>Time</td></tr>
<tr><td>Engage</td><td>• Think! If you freeze fruit salad, is it still a mixture?</td><td>TB p. 96</td><td>10 min</td></tr>
<tr><td>Explain</td><td>• Mixtures
• Got it? 60-Second Video (ActiveTeach)</td><td>SB p. 96–97
TB p. 97</td><td>40 min
10 min</td></tr>
<tr><td>Elaborate</td><td>• Mixture for Soap Bubbles
• At-Home Lab: Food We Eat at Home</td><td>TB p. 96
SB p. 97</td><td>20 min
20 min</td></tr>
<tr><td>Evaluate</td><td>• Lesson 3 Check (ActiveTeach)
• Assessment for Learning
• Review (Lesson 3)
• Got it? Self Assessment (ActiveTeach)
• Got it? Quiz (ActiveTeach)</td><td>TB p. 99a
TB p. 97
SB p. 99
TB p. 99b
TB p. 99b</td><td>10 min
10 min
10 min
10 min
10 min</td></tr>
<tr><td>Lab</td><td>• Let's Investigate! What is in a mixture? (ActiveTeach)</td><td>SB p. 98</td><td>30 min</td></tr>
</table>

Flash Cards

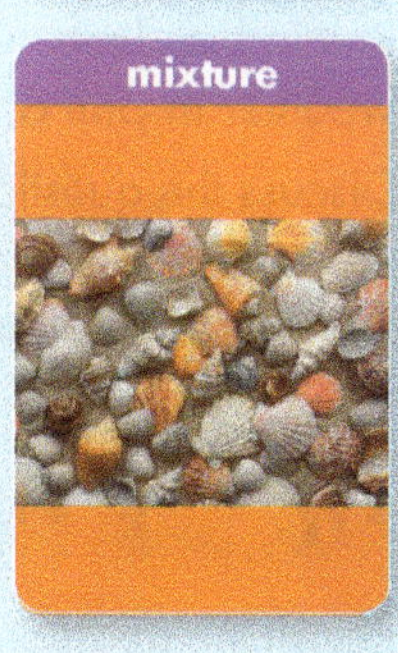

Lesson 1

Key Words	ELL Support
matter, solid, liquid, container, gas	**Vocabulary:** made of, wood, metal, plastic, wool, kind of, shape, move, building block, size, fish bowl, straw, balloon, fill, air, breathe, blow up, burst

Lesson 2

Key Words	ELL Support
freeze, ice, melt, boil	**Vocabulary:** change, paper, cut, modeling clay, mold, stretch, fold, half, tie, shoelace, warm, hot, snowman

Lesson 3

Key Words	ELL Support
mixture	**Vocabulary:** salad, soup, sand, shell, crayon **Vocabulary:** vegetables: tomato, carrot, cucumber, lettuce, cabbage, broccoli

Unit 8 — Matter and Mixtures

Unit Objectives

Lesson 1: Students will learn about solids, liquids, and gases.

Lesson 2: Students will learn how water can change.

Lesson 3: Students will learn about mixtures.

Vocabulary: *matter, shells, solid, liquid, container, gas, freeze, ice, melt, boil, mixture*

Introduce the Big Question

What are matter and mixtures?

Build Background Read the Big Question aloud and pre-teach *matter* with the *matter* Flash Card. Hold up and point to different objects in the classroom. Elicit what they are and what they are made of. (Possible answer: *It is a pen. It is made of plastic.*) *All of these things are kinds of matter, too. We are going to learn more about matter.*

Engage

Is this a mixture?

Draw students' attention to the picture on the lower right of the page. Identify the shells and read the question aloud. *Are the shells all mixed together?* Allow students time to discuss whether they think this is a mixture.

 **Circle the object that is strong and solid.**

Point to the pictures and have students describe what they see. *What material is the jug made of? Glass. Can you see through glass? Yes! What do you see in the second picture? Clouds and sky. There is air in the sky. Can you see the air? No! What material is the block made of? Wood. What is wood like? Strong!*

Have the students circle the object that is strong and solid. Check answers as a class. Explain to students that these pictures show different kinds of matter and that students will learn more about kinds of matter in this unit.

2 Look at the pictures. After the boy cuts shapes, are they still paper? Talk as a class.

Invite students to look at the first picture. *What is in the picture? Paper and scissors. What colors do you see?* Then have students describe what the boy is doing in the second picture. *He is cutting paper!*

Ask the question and discuss as a class. Help students understand that the paper can change shape, but it's still paper, even when we cut it.

3 Circle what you can mix in a bowl and eat for breakfast.

Focus students' attention on the pictures on the lower left of the page. Identify them with the students: *juice, cereal, milk. What do you mix in the bowl, juice and milk? Juice and cereal? Milk and cereal?* Have students circle the correct pictures and check as a class.

Elaborate

What are clouds made of?

Ask the question and have students discuss in groups. Remind students about what they learned about the weather in Unit 6, Lesson 3.

Discuss the question as a class. Elicit or explain that clouds are made of very small drops of water or pieces of ice. *The drops are so small they can float in the air.*

Think! Again!

Revisit the question *Is this a mixture?* Invite students to share their ideas freely. Accept all logical answers. *It's a mixture! There are different kinds of shells mixed together. But, there is only one thing in it, shells. It is not a mixture!*

What are solids, liquids, and gases?

Objective: Learn what matter is.

Vocabulary: *matter, made of, wood, metal, plastic, wool, kind of, solid, liquid, gas*

Digital Resources: Flash Cards (*matter, solid, liquid, gas*)

Materials: pictures of solids, liquids, and objects containing gases from magazines or the Internet, building blocks (enough for each student)

Unlock the Big Question

Write the following on the board: *I will learn about matter. What is it and what are the main kinds of matter? We are going to learn how matter can be a solid, liquid, or gas.*

Build Background Spread out the magazine pictures on a table. Ask students to come up and take a picture. Have them sit down in small groups and show one another their pictures. *What can you see in your picture? What is it made of? What is it like?* Have students discuss the questions in their groups. Monitor and provide support as necessary.

Explain

1 Read. What are two examples of matter? Say with a partner.

What is matter? Elicit ideas from the class. Then read the paragraph aloud and revisit the question using the *matter* Flash Card. Check comprehension by pointing to items and body parts and asking *Is this matter? Yes!*

Then display the Flash Cards or point to the pictures on the page and read the captions to help students understand three kinds of matter: solids, liquids, and gases. Explain that the air inside the balloon is a gas. Then put the students in pairs and have them think of two examples of matter. Elicit ideas as a class.

2 Look at the pictures of the juice and the fruit. Are they liquids or solids? Say with a partner.

Focus students' attention on the photos of the two girls. *What are they doing? She is drinking juice. She is eating fruit.* Help students compare the juice and fruit with the pictures of solids and liquids above. *Is the juice liquid or solid? Liquid. Are the fruits liquids or solids? Solids. Solids and liquids are both kinds of matter. They are different kinds of matter.*

Lesson 1 · What are solids, liquids, and gases?

1 Read. What are two examples of matter? Say with a partner.

Key Words
- matter
- solid
- liquid
- container
- gas

Matter

Matter is all around us. Matter is what all things are made of. The book that you are reading is made of matter. The air around you is made of matter. Materials like wood, metal, plastic, and wool are matter. There are different kinds of matter. Three kinds of matter are solids, liquids, and gases.

A building block is a solid.

Milk is a liquid.

The air in bubbles is a gas.

2 Look at the pictures of the juice and the fruit. Are they liquids or solids? Say with a partner.

Unit 8 **89**

Think!

Can you make a house out of juice?

Remind students what they've learned about choosing materials based on their properties. Then ask the question. Encourage students to explain their reasoning. Accept all logical answers.

Elaborate

What's the matter?

Clear space in the classroom and tape the *solid, liquid,* and *gas* Flash Cards to the walls around the classroom. Invite students to stand up and run to the type of matter when you call out a substance. For example, when you call out *Juice*, students run to the liquid sign. When you call out *What's inside a beach ball*, students run to the gas sign. Continue using several examples for each state of matter.

Building Blocks

Divide the class into small groups and distribute the building blocks. Have students create different shapes with the building blocks, e.g., squares or rectangles, by placing them on their tables. Then have students create a tower or other structure with their building blocks. Invite groups to present their creations to the class. Then have students disassemble their creations and sort the blocks by color. Ensure that students understand that, even though they can make different shapes with their building blocks, each block keeps its own shape and color.

What are solids, liquids, and gases?

Objective: Learn about solids.

Vocabulary: *solids, shape, move, building block, size*

Digital Resources: Flash Card (*solid*)

Materials: building blocks (a few for each student), some sugar, salt, and flour, hand lens

Build Background Show the *solid* Flash Card and elicit what is pictured. Show the building blocks and have students identify the colors. Then ask *Are they solids or liquids? Solids.* Distribute a few building blocks to each student and ask them to make something. Invite volunteers to show what they created to the class.

Explain

3 **Read. What is a solid? Say with a partner.**

Read the paragraph aloud with students. *What is a solid?* Elicit answers. (Possible answers: *It's a kind of matter. You can move it, and it keeps the same shape.*) Focus students' attention on the two pictures and read the captions aloud.

Then ask questions to reinforce the concept. *Look at the boy's hat? If he takes off his hat, is it still the same shape? Yes! Look at your desk. If we turn your desk around, is it the same shape? Yes! Move your pencil around your desk. Does it change shape? No!*

4 **Draw two solids in the classroom. What are they?**

Point to the two blank spaces and have students draw two solids they can see in the classroom. Monitor and provide support. If students are having difficulty, remind them of the examples in their books or the examples you asked them about in the previous exercise.

When they finish, have students show their drawings and identify the solids in small groups.

5 **What are the solids in your drawings like? What color are they? What size are they? What shape are they? Talk with a partner.**

Read the questions aloud. Draw a classroom object in color on the board and demonstrate how to answer the questions.

Have the students talk about their drawings in pairs. Help students identify the properties of the items they drew. (Possible answers: *It is (green). It is a (square). It is a (solid).*) Monitor and provide support as needed.

---

3 **Read. What is a solid? Say with a partner.**

Solids

Solids are one kind of matter. **Solids** keep their shape. A table is a solid. You can move the table. It still keeps its shape. You can make a house with building blocks. The blocks keep their shape.

The table is a solid.

The building blocks are solids.

4 **Draw two solids in the classroom. What are they?**

5 **What are the solids in your drawings like? What color are they? What size are they? What shape are they? Talk with a partner.**

90   Unit 8

Elaborate

Sugar, salt, and flour: are they solids?

Bring in small amounts of sugar, salt, and flour in separate pieces of foil. Open up the foil packages on a table and have students touch the sugar, salt, and flour. Identify them and discuss what they are used for. Provide vocabulary support as necessary.

Are they solids? Have students discuss as a class. Invite them to look at the sugar, salt, and flour through a hand lens. Explain that sugar, salt, and flour are like tiny building blocks. *They keep the same shape and size when you move them. They are solids!*

Think!

Is sand a solid?

Ask the question and have students discuss in groups. If necessary, help them by asking *Is sand like sugar, salt, and flour? Yes! Right! The grains of sand are like tiny building blocks, too. They are solids!*

What are solids, liquids, and gases?

Objective: Learn about liquids.

Vocabulary: *liquid, shape, fish bowl, container, solid, straw, pour*

Digital Resources: Flash Cards (*liquid, lake, river*), I Will Know... Digital Activity (*Optional:* Unit 2 *I Will Know...* Digital Activities)

Materials: bottle of water, large glass, jar of liquid honey (ideally with an interesting shape), spoon, clear drinking straw (ideally with an interesting shape), jug or beaker of colored water, fish bowl

Build Background Show the *liquid* Flash Card. *What is it? Water. Where can you find water on Earth?* Remind students about what they learned in Unit 6, Lesson 1.

Explain

6 **Read. Is water a liquid? Say with a partner.**

Read the paragraph aloud and focus students' attention on the two pictures. *Do the container and the fish bowl have the same shape? No. Does water take their shape? Yes. Is water a liquid? Yes.* Display the *lake* Flash Card. *What shape does the lake water take?* Elicit or explain that the lake takes the shape of the land around it. Repeat with the *river* Flash Card.

To consolidate understanding, put the fish bowl on a table. *What is the fish bowl made of? Glass! Can you see through the glass? Yes!* Show students the container of colored water and explain that you are going to pour some of the water into the fish bowl. *What happens to the water? Does it take the shape of the fish bowl? Yes! What kind of matter is it? Liquid.* Then move the fish bowl to another table. *Does the fish bowl keep the same shape? Yes! What kind of matter is it? A solid!*

7 **Look at the picture of the fish bowl. What is solid? What is liquid? Say as a class.**

Draw students' attention to the picture of the fish bowl again. Read the questions aloud and elicit the answers. (Possible answers: *The goldfish is solid. The fish bowl is solid. The water is liquid.*)

8 **What are some other liquids? Make a list with a partner.**

Show the jar of honey. Take spoonfuls and let the honey pour back into the jar. *Watch while I pour the honey back into the jar. What shape does it take? The shape of the jar! Is the honey liquid? Yes.* Have students list other liquids in pairs. Monitor and provide support. Elicit ideas from the class. (Possible answers: *milk, juice, water in lakes*)

6 Read. Is water a liquid? Say with a partner.

Liquids

Liquids are a kind of matter. **Liquids** take the shape of their **containers**. Water is a liquid. You can put water into a fish bowl. It takes the shape of the fish bowl.

The liquid takes the shape of its container.

The water takes the shape of the fish bowl.

7 Look at the picture of the fish bowl. What is solid? What is liquid? Say as a class.

8 What are some other liquids? Make a list with a partner.

9 Color a liquid in this straw. What shape does the liquid take?

9 **Color a liquid in this straw. What shape does the liquid take?**

What do straws help us do? They help us not spill. Have students color a liquid in the straw. Read the question. Elicit or explain that the liquid takes the shape of the straw. Show the *I Will Know...* Digital Activities from Unit 2 to help reinforce understanding.

Elaborate

⚡ Flash Lab

The Shape of Water

Do the Lab. Then ask what happens if you pour too much water. *It spills. Right! Water takes the shape of its container. If you pour too much, it spills over the sides!*

How does a straw work?

Suck some colored water into the straw and seal it with your finger. Show students that the water is still inside it. Then let go and show how the water falls out of the straw into the glass. Discuss how the water changed shape.

I Will Know...

Have students do the *I Will Know...* Digital Activity.

What are solids, liquids, and gases?

Objective: Learn about gases and how they are different from liquids.

Vocabulary: *gases, matter, shape, container, balloon, fill, air, breathe, blow up*

Digital Resources: Flash Cards (*liquid, gas*), *Lesson 1 Check* (print out 1 per student)

Materials: balloons (different shapes, if possible)

Build Background Blow up a balloon slowly in class. *What is this? A balloon. How do you blow up a balloon?* Elicit or explain that you blow air inside it to fill it up. Tap the balloon softly and have students play with it for a few seconds. Discuss how the balloon feels like it could burst.

Explain

10 Read. How are gases different from liquids? Say with a partner.

Show the *liquid* and *gas* Flash Cards. Invite students to say what they know about liquids. Read the paragraph aloud. Draw students' attention to the balloons. *What is inside the balloons? Air.* Turn to the pictures of liquids on page 91. Point to the line where the liquid stops in each. *Liquids take the shape of their containers, but they don't fill up the whole container like gases do.*

How are gases different from liquids? Have students answer in pairs and check answers as a class. (Possible answer: *Gases fill the whole container.*)

11 Which of these things are filled with gas? Circle.

Focus students' attention on the three pictures and elicit or identify what they are: *beach ball, waterwings,* and *soap bubbles.* Elicit or explain how each thing is used. (Possible answers: *You blow up a beach ball and waterwings. You can play with the ball. The waterwings help you float. You make bubbles with the bubble wand.*) *Right! You put air inside all of these things! The air fills them up!* If necessary, remind students that air is a gas. Then have students circle the things that are filled with gas. Check answers as a class.

12 Look at the balloons. What happens if you blow them up? Say as a class.

Draw students' attention to the picture of the balloons. Ask the question and elicit answers from the class. Inflate a balloon and hold it up. *Air fills the balloon. Gas takes the shape of its container.*

---

10 Read. How are gases different from liquids? Say with a partner.

Gases

Gases are a kind of matter. Gases take the shape of their containers, too. They fill the whole container! Balloons can be filled with gas. The gas fills the whole balloon. The air we breathe is a gas. It is all around us.

Gas fills these balloons!

11 Which of these things are filled with gas? Circle.

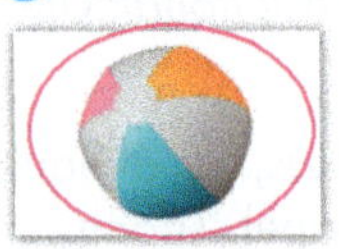

12 Look at the balloons. What happens if you blow them up? Say as a class.

13 Look around the classroom. Are there things filled with gas? Talk as a class.

---

13 Look around the classroom. Are there things filled with gas? Talk as a class.

Read the question aloud and invite volunteers to find things that are filled with gas. It's unlikely there will be many. Have students compare how many solids they can point to versus liquids and gases. Elicit or explain that you can't see most gases. *Sometimes, like the balloon, you can only see the container.*

Elaborate

The Sounds of a Balloon

Blow up a balloon and open its mouth a little bit. Let the air out slowly to make a funny noise. *What is making this sound?* Discuss that it's the air coming out of the balloon. Blow the balloon up again and hold its mouth closed. Ask students what will happen if you let go. Elicit ideas. Let go so that students can see the balloon fly around the classroom. Challenge students to consider what happened to the gas inside the balloon. Guide students to understand that it came out of its container, the balloon, and filled up the classroom!

Evaluate

Lesson 1 Check Assessment for Learning

Review the Key Words for Lesson 1 (see Student's Book page 89). Distribute the *Lesson 1 Check* and guide students as they complete it. Check answers as a class. Then ask students to grade their progress on the topic of solids, liquids, and gases from 1 to 3: 3 = *I understand what solids, liquids, and gases are;* 2 = *I need to study more;* 1 = *I need help!* Encourage students giving themselves a 1 or a 2 to say what they found difficult and what they need to study more.

How can matter change?

> **Objective:** Learn about some ways matter can change.
>
> **Vocabulary:** *matter, change, paper, cut, size, modeling clay, mold, stretch, fold, half, tie, shoelace*
>
> **Digital Resources:** Flash Cards (*liquid, ice*), *Let's Explore!* Digital Lab, Origami Frog Resource
>
> **Materials:** sheets of paper cut in half, safety scissors, modeling clay (some for each student), shoelaces, green paper (1 per student)

Unlock the Big Question

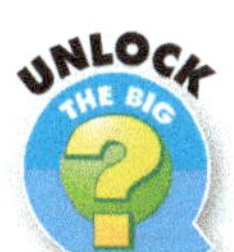

Write the following on the board: *I will learn some ways matter can change.*

Build Background Display the *liquid* Flash Card. *What's inside the glass?* Water. *Is it a liquid or a solid?* Liquid. Then display the *ice* Flash Card. *Is it a liquid, a solid, or a gas?* Invite volunteers to say, but don't confirm answers yet.

Explore

Let's Explore! Lab How do materials change?

Objective: Learn how paper and modeling clay can change.

Digital Resources: *Let's Explore!* Digital Lab, *Let's Explore!* Activity Card (1 per student) (*Optional*: Show the Digital Lab instead.)

- Display the materials on a table and have students identify them: *paper, scissors, modeling clay.*

- Hand out a sheet of paper and a pair of scissors to each student. Ask them to copy what you do. Cut the sheet of paper in three pieces. Then take one piece and cut it into small pieces. Crumple the second piece and tear the third one. *How did the paper change? Is it still paper?* Distribute the *Activity Card* and ask students to draw pictures of the paper.

- Distribute the modeling clay. Scrunch, stretch, and twist it. Each time, have students do the same and draw pictures on their *Activity Cards.*

Explain

1 **Read. You make a shape with modeling clay. Is it still modeling clay? Talk with a partner.**

Mold the clay and divide it into different shapes. *This is a (triangle). Is it still modeling clay?* Read the paragraph aloud and have students check their answers. Focus their attention on the picture of

---

> **Lesson 2 · How can matter change?**
>
> **1** **Read.** You make a shape with modeling clay. Is it still modeling clay? Talk with a partner.
>
> **Key Words**
> - freeze
> - ice
> - melt
> - boil
>
> **Properties of Matter Can Change**
>
> Matter can change. Paper is a kind of matter. You can cut paper into different sizes. It is still paper. Modeling clay is also a kind of matter. You can mold modeling clay into different shapes. You can stretch it. It is still modeling clay. It is the same kind of matter.
>
> 
> *You can make numbers with the modeling clay!*
>
> **2** **Fold** a piece of paper in half. Then fold it again. Is it still paper? Talk with a partner.
>
> **3** **You can tie your shoelace. Is it still a shoelace? Talk with a partner.**
>
> shoelace
>
> Let's Explore! Lab   Unit 8   93

the numbers made with modeling clay. Distribute modeling clay to students and invite them to mold it into a number. *Are your clay numbers still made of clay?* Yes!

2 **Fold a piece of paper in half. Then fold it again. Is it still paper? Talk with a partner.**

Distribute the pieces of paper. *What is this?* Paper. Show students how to fold the paper in half and then fold it again. Allow partners to freely discuss.

3 **You can tie your shoelace. Is it still a shoelace? Talk with a partner.**

Show the shoelaces and pre-teach *shoelace*. Check which students are wearing shoes with shoelaces. Tie the shoelaces and ask the question. Elicit that they are still shoelaces. Guide students to conclude that we can change the size and shape of matter (by cutting, folding, molding, etc.), but it is still the same kind of matter.

Elaborate

Origami Frog

Show a sheet of paper. *What can you make with paper? How does it change?* Discuss with students.

Then show them an origami frog (or other animal or shape) that you have made previously. Hand out sheets of paper and help students make the shape themselves. Discuss how many times you folded and unfolded the paper. Then ask *Is it still paper? Can you write on it? Can you unfold it?* Yes!

How can matter change?

> **Objective:** Learn how water can change.
>
> **Vocabulary:** *freeze, ice, water, solid, liquid, gas, melt, hail*
>
> **Digital Resources:** Flash Cards (*liquid, freeze, ice*), *I Will Know...* Digital Activity
>
> **Materials:** jug, ice cube tray, water glass, water bottle, watering can (all empty)

Build Background Display the objects on a table. Put the students in small groups and ask them to find what is common to these objects. Elicit answers as a class. (Possible answer: *You can put water in them.*) Discuss what each object is used for. Provide support as necessary.

Explain

4 **Read. How does water change into ice? Say with a partner.**

Read the paragraph aloud and display the *liquid* and *ice* Flash Cards. *Which is the liquid water? Which is the frozen water? Which is the solid water?* Have students point to the correct Flash Cards.

Draw students' attention to the pictures on the upper right of the page. Read the captions aloud. Reinforce that water can be solid, liquid, or gas. Point to the picture of the ice sculpture. *Look! Someone made shapes out of frozen water! Is it still water? Yes!*

5 **Mark (✔) the picture that shows solid water.**

Read the instructions aloud and have students mark the correct picture, by process of elimination if necessary. Check answers as a class. Activate prior knowledge by asking *What kind of water falls from the sky? Rain! Right!* Explain that the first picture shows another kind of water that falls from the sky. *It's called hail. It's like frozen rain.*

6 **Water can change from a liquid to a solid. It can ________.**

Display the *melt* Flash Card and explain that the upper part of the water is frozen and the lower is changing from a solid (ice) into a liquid (water) to pre-teach *melt*. Show the *freeze* Flash Card and ask *Do popsicles melt? Yes. Do you think frozen water can melt? Yes. Is it still water? Yes!*

Read the instruction aloud and have students choose the correct word to complete the sentence. Have them write it on the line.

Think!

Where can water freeze?

Ask the question and have students discuss in small groups or as a class. Elicit possible answers. (Possible answers: *in the freezer, in rivers, or in lakes*) *What's the weather like when lakes freeze?* (Possible answer: *It's very cold outside.*)

Think!

Look at the picture. Does the liquid change?

Draw students' attention to the picture on the lower right of the page. Explain that it's an experiment with colored water. Read the question and have students discuss in pairs. Elicit answers as a class. (Possible answer: *The yellow and blue colored water change into green water when they are mixed.*) *Does the colored water take the shape of the new container? Yes!*

> **I Will Know...**
>
> Have students do the *I Will Know...* Digital Activity.

How can matter change?

> **Objective:** Learn that water can melt and boil.
>
> **Vocabulary:** *melt, boil, ice, warm, solid, liquid, hot, gas*
>
> **Digital Resources:** Flash Cards (*freeze, melt, boil, ice, liquid, solid, gas*), *Lesson 2 Check* (print out 1 per student)
>
> **Materials:** portable camping stove, gas, lighter, small pot, bottle of water, or short video in which you can see and hear boiling water

Build Background Write *Water* on the board and have students brainstorm what they know about it. If necessary, show them the *freeze* and *melt* Flash Cards to help them add to their ideas. If they haven't mentioned *boil*, help them think about it. *What do we do to water in the kitchen?* Don't check answers yet.

Explain

7 Read. How does ice change into water? Say with a partner.

Read the paragraph aloud. Then read the question aloud and elicit answers from the class. (Answers: *It gets very cold. It freezes.*)

Next, draw students' attention to the pictures labeled *melt* and *boil*. *In which picture is water cold?* (*In the first picture.*) *In which is it hot?* (*In the second picture.*) *Look! You can see bubbles in the water! The bubbles are made of gas. You can see gas bubbles in the water when it boils.*

Sum up all the ways in which water can change: *It can freeze, melt, and boil.*

8 Look at the sentences. Write.

Display the *ice, melt,* and *boil* Flash Cards on one side of the board. Display the *liquid, gas,* and *solid* Flash Cards on the other side of the board. Invite volunteers to match Flash Cards on the left and Flash Cards on the right. Have students explain their reasoning.

Then have students read the sentences and complete them with the words. Monitor and provide support.

Elaborate

The Sound of Boiling Water

What does boiling water sound like? Elicit answers from the class. Ask the students to remain seated and quiet throughout the experiment. Put some water in the pot and

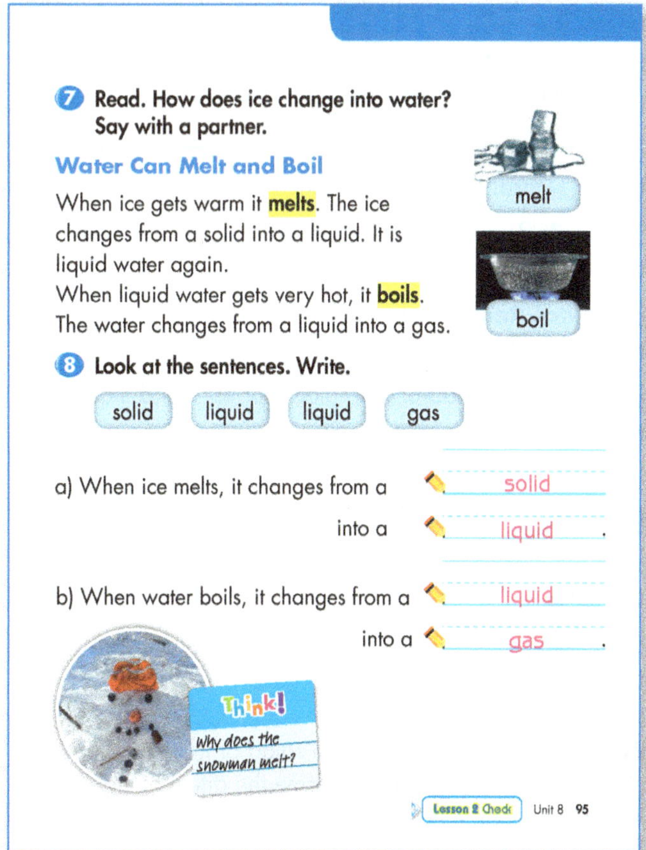

light the camping stove. Wait until the water is boiling with big bubbles. Alternatively, show a video clip of water boiling. Have students listen and imitate the sound. *Can you see bubbles in the boiling water? Yes! What do you think is inside the bubbles? Gas! What kinds of matter are there? Liquid (water), gas (bubbles), and solid (pot).*

Think!

Why does the snowman melt?

Draw students' attention to the picture of the snowman. Elicit or explain what it is and elicit whether students have ever made a snowman or other shape with snow or ice. Discuss how they made it and what materials they used. Then ask *Did your snowman melt? When?* Discuss that snowmen melt when it gets hot outside.

Evaluate

Lesson 2 Check Assessment for Learning

Review the Key Words for Lesson 2 (see Student's Book page 93). Distribute the *Lesson 2 Check* and guide students as they complete it. Check answers as a class. Then ask students to grade their progress on the topic of how matter changes from 1 to 3: 3 = *I understand how matter changes;* 2 = *I need to study more;* 1 = *I need help!* Encourage students giving themselves a 1 or a 2 to say what they found difficult and what they need to study more.

What is a mixture?

> **Objective:** Learn what a mixture is.
>
> **Vocabulary:** *mixture, salad, vegetables, soup, solid, liquid, gas*
>
> **Digital Resources:** Flash Card (*mixture*), *I Will Know...* Digital Activity
>
> **Materials:** salad or fruit salad, bubble wand, liquid soap, water

Unlock the Big Question

Write the following on the board: *I will learn about mixtures.*

Build Background Show the salad or fruit salad and elicit what it is. Help students name the different ingredients. *Yes! All of these things are put together to make the salad. We call the salad a mixture.*

Explain

1 **Read. Look at the picture of the salad. Mark (✔) the food that is in the salad.**

Read the paragraph aloud for students. Check comprehension by asking *Did you have a mixture for breakfast?* Elicit ideas, e.g., *bowl of cereal with milk, stew, soup.*

Draw students' attention to the picture of the soup and elicit what's in it. Then ask them to look at the picture of the salad and mark the pictures of the individual vegetables that make up the salad. Check answers as a class.

ELL Vocabulary Support

Review or teach the words for the vegetables in the pictures: *tomato, carrot, cucumber, lettuce, cabbage, broccoli.* Have students label the pictures in their books or draw and label the pictures in their notebooks.

2 **What is in the soup? Are there solids, liquids, or gases? Say with a partner.**

Read the question aloud and draw students' attention to the picture of the soup again. Have students discuss their answers in pairs. Check as a class. (Possible answers: *There are vegetables and water. There are liquids and solids.*)

Ask *What is in the salad? Solids, liquids, or gases?* Elicit answers. (Possible answers: *solids and liquids, like oil and vinegar*)

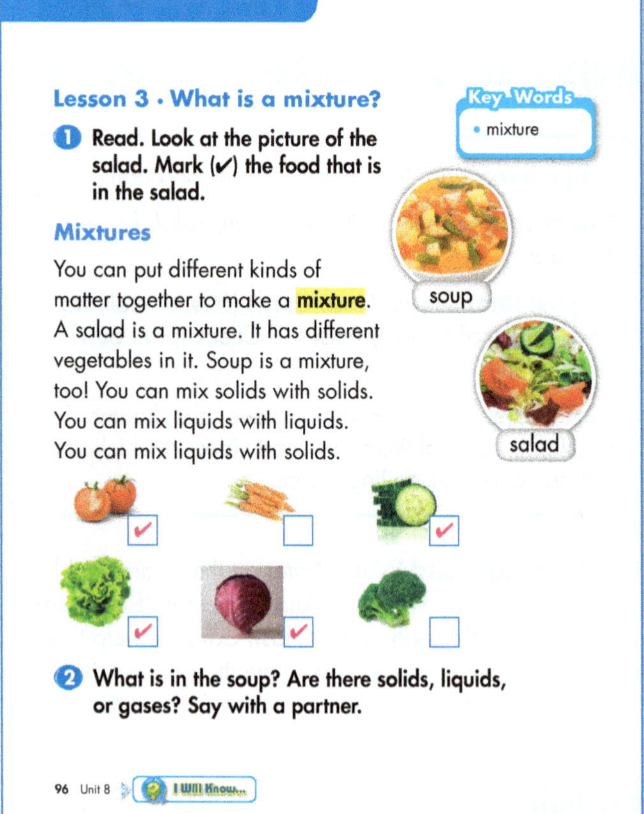

Think!

If you freeze fruit salad, is it still a mixture?

Read the question aloud for students. Monitor and provide support. Guide students to conclude that, even though it is frozen, it is still a mixture because there are different kinds of fruit mixed together.

Elaborate

Mixture for Soap Bubbles

Hold up the bubble wand. *What's this? Do you have one? What can you make with it?* Elicit answers from the class. Then ask *Do you use a mixture with it? What do you make the mixture with?* Elicit ideas and show the water and liquid soap.

Make a bubble solution by mixing water and liquid soap. Blow a few bubbles and ask *Are there solids, liquids, or gases in the mixture?* (Answer: *There are liquids: water and liquid soap.*) *Are there solids, liquids, or gases inside the bubbles?* (*Gas, the air we blow to make the bubbles.*) Invite volunteers to blow bubbles and/or catch them. Challenge students to answer what happens to the gas inside the bubbles when they burst. *It goes into the air around us!*

> **I Will Know...**
> Have students do the *I Will Know...* Digital Activity.

What is a mixture?

Objective: Learn more about mixtures.

Vocabulary: *mixture, matter, liquid, solid, gas, sand, shell, crayon*

Digital Resources: Flash Cards (*freeze, mixture*), *Lesson 3 Check* (print out 1 per student), *Got it? 60-Second Video*

Materials: container with shells and sand

Build Background Display the *freeze* Flash Card. *What are they?* (Popsicles.) *Are they solid, liquid, or gas?* (Solid.) *Can they change into liquid?* (Yes. They can melt.)

Explain

3 **Some mixtures have two kinds of matter. Think of some examples. Talk as a class.**

Read the instructions. *What is a mixture that has a solid and a liquid?* (Possible answers: *soup with vegetables or meat*) *What mixture has liquid and gas?* (Possible answers: *soda pop*) *What mixture has a solid and a gas?* (Possible answer: *soda pop with ice*).

4 **Look at the picture. What kinds of matter are there? Talk as a class.**

Invite volunteers to say what's in the glass. *Liquid (soda pop), gas (bubbles in the soda pop), and solids (ice cubes).*

5 **Look at the picture of sand and shells. Is it a mixture?**

Draw students' attention to the picture of the sand and shells. Ask the question and elicit that the sand and shells are a mixture. Encourage students to explain their reasoning. (*It is a mixture. It has two things mixed together, sand and shells.*)

6 **Pretend you put the shells and sand in different groups. Is there still a mixture? Talk as a class.**

Read the instructions aloud and, if possible, have students look at a mixture of real shells and sand. Invite students to take out the shells. Accept all logical answers. (Possible answers: *No, we made two groups, one of sand and one of shells*).

7 **Look at the picture. Are the crayons a mixture? Why or why not? Talk in small groups.**

Read the questions aloud and allow students to discuss freely. Accept all logical answers. (Possible answers: *They are different colors. They are a mixture! They are all crayons. They are not a mixture!*)

---

3 Some mixtures have two kinds of matter. Think of some examples. Talk as a class.

4 Look at the picture. What kinds of matter are there? Talk as a class.

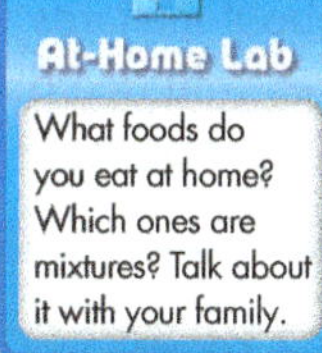

5 Look at the picture of sand and shells. Is it a mixture?

6 Pretend you put the shells and sand in different groups. Is there still a mixture? Talk as a class.

7 Look at the picture. Are the crayons a mixture? Why or why not? Talk in small groups.

Lesson 3 Check | Got it? | 60-Second Video | Unit 8 97

Elaborate

At-Home Lab

Food We Eat at Home

Explain the exercise to students and have them give an example in class the next day.

Evaluate

Lesson 3 Check Assessment for Learning

Review the Key Words for Lesson 3 (see Student's Book page 96). Distribute the *Lesson 3 Check* and guide students as they complete it. Check answers as a class. Then ask students to grade their progress on the topic of mixtures from 1 to 3: 3 = *I understand what mixtures are*; 2 =*I need to study more*; 1 = *I need help!* Encourage students giving themselves a 1 or a 2 to say what they found difficult and what they need to study more.

Got it? 60-Second Video
Play the *Got it? 60-Second Video* to review the unit material.

Let's Investigate!

In this unit, students learn about how matter changes and about mixtures. In this lab, they will create their own mixtures with rice and beans and sort them into their component parts.

Let's Investigate! Lab **What is in a mixture?**

Objective: Students will create and sort mixtures with rice and beans.

Materials: one set per student: sheet of white paper, rice, black beans, red beans (5 of each), transparent cup, spoon

Digital Resources: *Let's Investigate!* Digital Lab, *Let's Investigate!* Activity Card (1 per student)

Advance Preparation: Prepare the rice and beans by placing five of each in separate bags. Each student will need a set of rice, black beans, and red beans.

- Explain that students are going to create mixtures with rice and beans. Show them the rice and beans. Distribute the sheets of paper, rice, and beans.

- Have the students place the rice and beans on the paper in separate groups. Ask them to count how many grains of rice, red beans, and black beans they have. Distribute the *Activity Cards* and have students write the numbers in the table.

- Then distribute the cups and spoons. Have students put the rice and beans in the cups and mix them with the spoons. Invite them to look at the cups and notice how the rice and beans have mixed.

- Have students pour the rice and beans onto the paper again. Ask them to sort them into groups and count each again. Have them write the figures in the table on the *Activity Card.*

- Discuss how there are the same number of rice grains and beans when they are in separate groups and when they are in the mixture.

Teacher Time-Saving Option: Show the *Let's Investigate!* Digital Lab as an alternative to the hands-on lab activity.

Unlock the Big Question

Have students refer to the Big Question on the Unit Opener page. In pairs have them recall what they have learned about matter and mixtures. Have pairs complete questions 4 and 5 on the *Activity Card.*

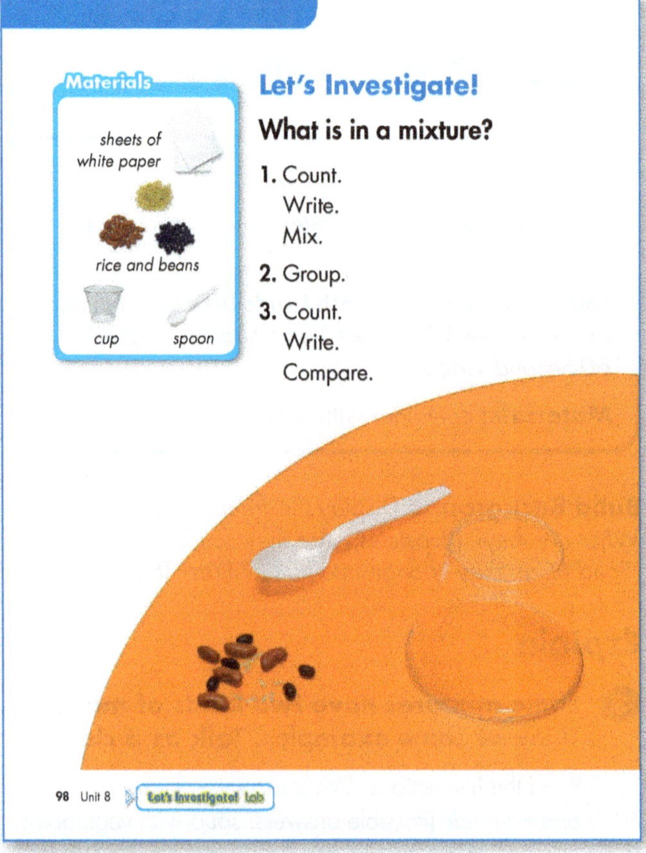

Class Project: Snowflake Collage

Materials: Snowflake Pattern Resource, sheets of white paper, safety scissors (1 pair per student), construction paper, glue

Prepare a paper snowflake before class. Show it to students and elicit or explain what it is. Explain that this is what a snowflake looks like through a microscope. *Is a snowflake liquid or solid?* (*Solid. It's frozen water.*) Then explain that each snowflake is different. *There are no two snowflakes that are identical, or the same.*

Hand out the sheets of paper and help students fold the paper to make snowflakes. Then hand out the scissors and have students cut out random shapes. Monitor and provide support as needed.

Students can unfold their paper to look at their snowflakes. Have them compare in small groups. Allow them to decorate their snowflakes by drawing pictures and/or coloring them.

Then glue the snowflakes on the large construction paper. Add a heading: *Snowflakes are...* Have students help you list some properties of snowflakes. *What kind of matter are they? Solids! Are they made of water? Yes! What kind of water? Frozen! What shape are they? They are all different shapes? What size are they? Small.* Display the poster in class.

Unit 8 Review

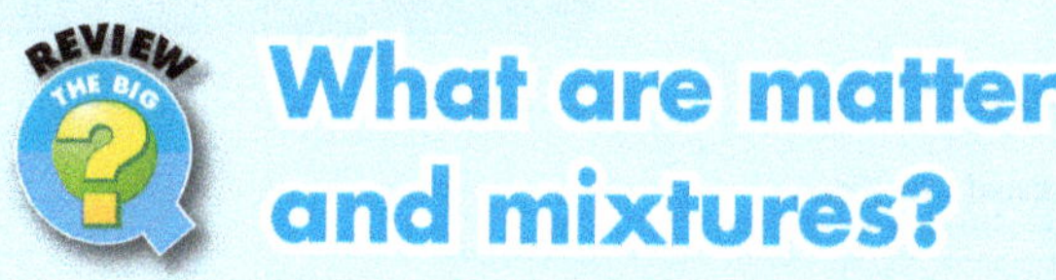

What are matter and mixtures?

Digital Resources: Print out 1 of each per student: *Got it? Self Assessment, Got it? Quiz*

Evaluate

Strategies for Targeted Review

The following are strategies for providing targeted review for students if they encounter challenges with the content.

Lesson 1 What are solids, liquids, and gases?

Question 1

If... students are having difficulty answering the question, then... direct students to review Lesson 1.

Lesson 2 How can matter change?

Question 2

If... students are having difficulty matching the sentences to the pictures, then... read the sentences aloud as a class.

Lesson 3 What is a mixture?

Question 3

If... students are having difficulty marking the picture of the mixture, then... explain that soup is a mixture. Help students extrapolate to select the salad.

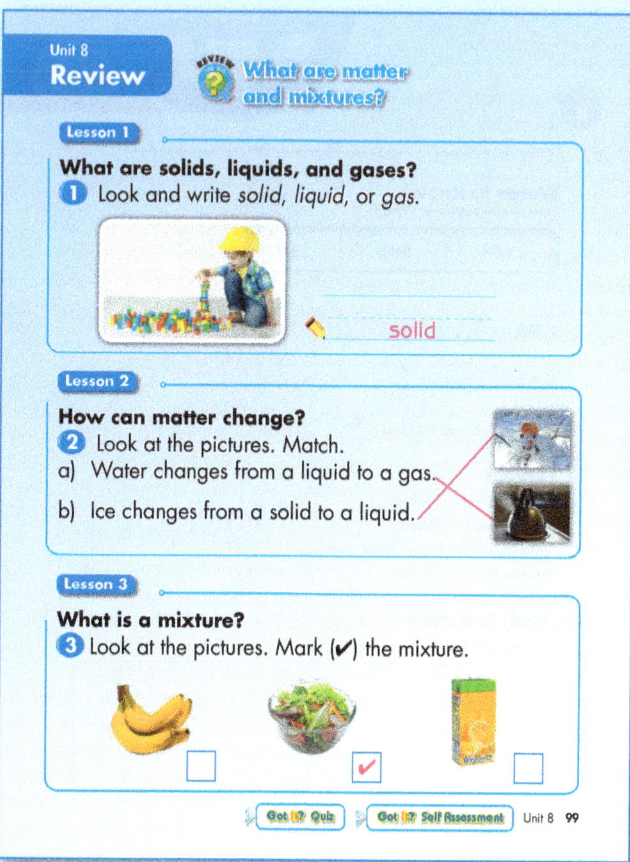

ELL Language Support

Before students start working on the Review activities, have them read each question aloud along with you.

Got it? Self Assessment

Immediately after students have completed the Review activities, distribute a *Got it? Self Assessment* to each student. Have students complete the *Stop! Wait!* and *Go!* statements for each lesson, allowing them to look back through the lesson material if necessary.

Got it? Quiz

Distribute a Unit 8 *Got it? Quiz* to each student. Quizzes may be used for assessing students' understanding of unit concepts as well as for grading purposes.

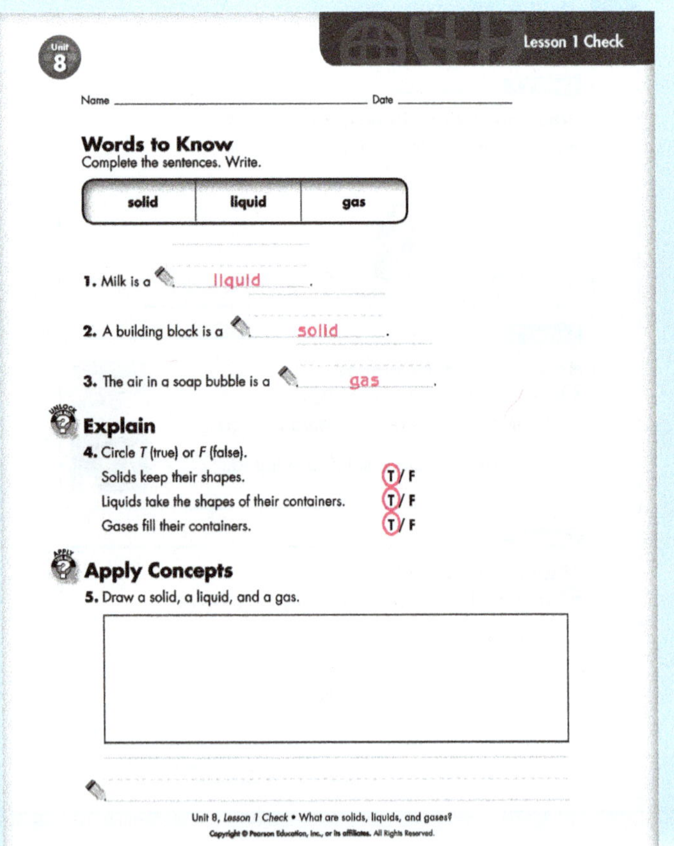

Words to Know

Complete the sentences. Write.

| solid | liquid | gas |

1. Milk is a __liquid__.

2. A building block is a __solid__.

3. The air in a soap bubble is a __gas__.

Explain

4. Circle *T* (true) or *F* (false).

Solids keep their shapes. (T)/ F

Liquids take the shapes of their containers. (T)/ F

Gases fill their containers. (T)/ F

Apply Concepts

5. Draw a solid, a liquid, and a gas.

Unit 8, *Lesson 1 Check* • What are solids, liquids, and gases?
Copyright © Pearson Education, Inc., or its affiliates. All Rights Reserved.

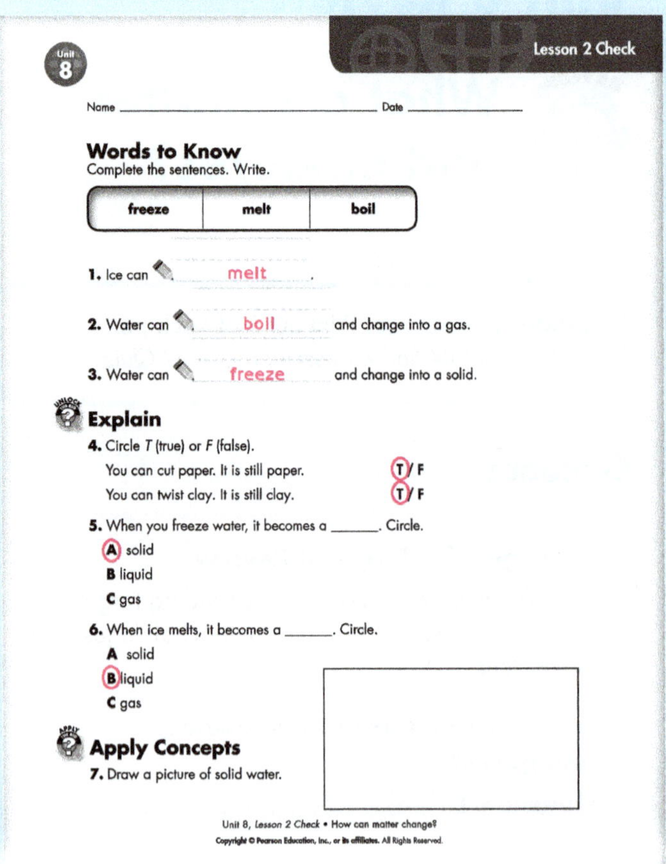

Words to Know

Complete the sentences. Write.

| freeze | melt | boil |

1. Ice can __melt__.

2. Water can __boil__ and change into a gas.

3. Water can __freeze__ and change into a solid.

Explain

4. Circle *T* (true) or *F* (false).

You can cut paper. It is still paper. (T)/ F

You can twist clay. It is still clay. (T)/ F

5. When you freeze water, it becomes a ______. Circle.

(A) solid

B liquid

C gas

6. When ice melts, it becomes a ______. Circle.

A solid

(B) liquid

C gas

Apply Concepts

7. Draw a picture of solid water.

Unit 8, *Lesson 2 Check* • How can matter change?
Copyright © Pearson Education, Inc., or its affiliates. All Rights Reserved.

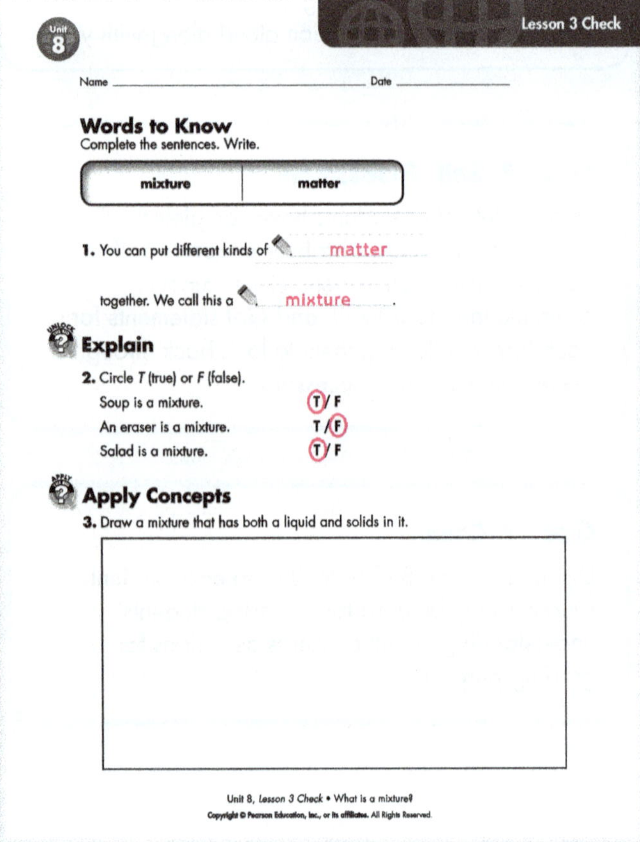

Words to Know

Complete the sentences. Write.

| mixture | matter |

1. You can put different kinds of __matter__

together. We call this a __mixture__.

Explain

2. Circle *T* (true) or *F* (false).

Soup is a mixture. (T)/ F

An eraser is a mixture. T /(F)

Salad is a mixture. (T)/ F

Apply Concepts

3. Draw a mixture that has both a liquid and solids in it.

Unit 8, *Lesson 3 Check* • What is a mixture?
Copyright © Pearson Education, Inc., or its affiliates. All Rights Reserved.

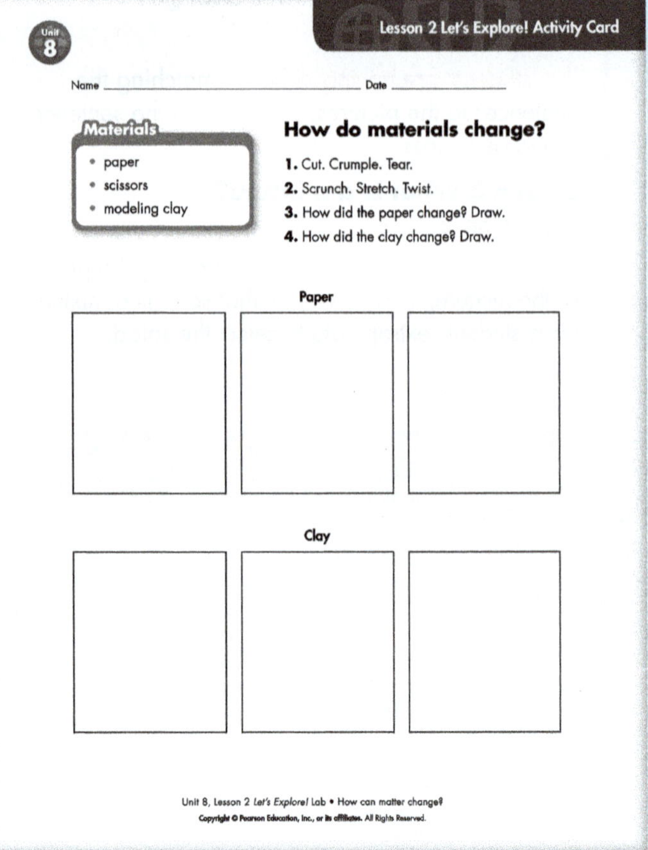

Materials

- paper
- scissors
- modeling clay

How do materials change?

1. Cut. Crumple. Tear.

2. Scrunch. Stretch. Twist.

3. How did the paper change? Draw.

4. How did the clay change? Draw.

Paper

Clay

Unit 8, *Lesson 2 Let's Explore! Lab* • How can matter change?
Copyright © Pearson Education, Inc., or its affiliates. All Rights Reserved.

Name _______________________ Date _______________

What is in a mixture?

4. Record.

Before Mixture		
Rice	Black Beans	Red Beans
5	5	5

After Mixture		
Rice	Black Beans	Red Beans
5	5	5

5. Compare.

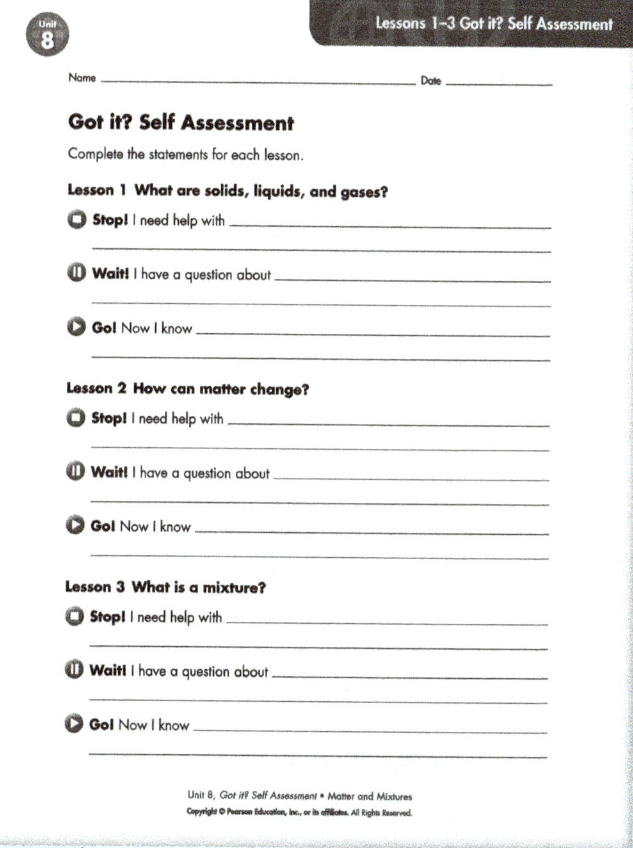

Name _______________________ Date _______________

Got it? Self Assessment

Complete the statements for each lesson.

Lesson 1 What are solids, liquids, and gases?

⬛ **Stop!** I need help with ____________________________
____

⏸ **Wait!** I have a question about ____________________

▶ **Go!** Now I know ________________________________
____

Lesson 2 How can matter change?

⬛ **Stop!** I need help with ____________________________
____

⏸ **Wait!** I have a question about ____________________

▶ **Go!** Now I know ________________________________

Lesson 3 What is a mixture?

⬛ **Stop!** I need help with ____________________________

⏸ **Wait!** I have a question about ____________________

▶ **Go!** Now I know ________________________________
____

Name _______________________ Date _______________

Got it? Quiz

1. Which sentence is true? Circle.
   A The air around us isn't matter.
   B Our science book is matter.
   C Matter never changes.

2. Solids __________. Circle.
   A take the shape of their containers
   B fill their containers
   C keep their shapes

3. Liquids __________. Circle.
   A take the shapes of their containers
   B fill their containers
   C keep their shapes

4. Gases __________. Circle.
   A take the colors of their containers
   B fill their containers
   C keep their shapes

Name _______________________ Date _______________

5. When you tear paper, it __________. Circle.
   A changes into clay
   B is still paper
   C changes into a gas

6. When water gets very cold, it __________. Circle.
   A melts
   B freezes
   C boils

7. When water gets very hot, it __________. Circle.
   A melts
   B freezes
   C boils

8. Which is a mixture? Circle.
   A ice
   B water
   C soup

Unit 8 • Digital Resources and Photocopiables **T99b**

Unit 8 Study Guide

What are matter and mixtures?

Lesson 1
What are solids, liquids, and gases?

- Matter is all around us.
- Solids, liquids, and gases are different kinds of matter.

Lesson 2
How can matter change?

- You can cut paper, but it is still paper.
- Water is a liquid that can become a gas or a solid.

Lesson 3
What is a mixture?

- You can put different kinds of matter together to make a mixture.
- A mixture can be made of different kinds of solids. It can be made of solids and liquids; It can be made of liquids and gases; It can be made of solids and solids; and so on.

Review the Big Question

What are matter and mixtures?

Have students use what they have learned from the unit to answer the question in their own words.

How has your answer to the Big Question changed since the beginning of the unit? What are some things you learned that caused your answer to change?

Make a Concept Map

Draw on the board a concept map like the one shown on this page. With the students, talk through the key ideas from this unit. Invite different students to point to the ideas on the board, miming as possible.

 Unit 8 Concept Map

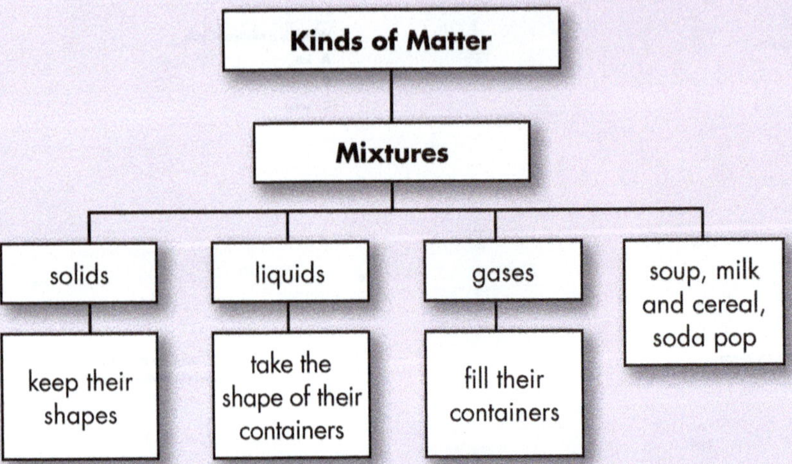

Students can make a concept map to help review the Big Question.

Unit 9 Motion

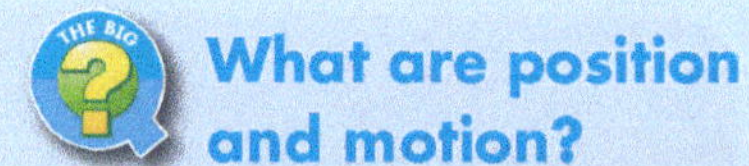

Lesson Plan

Unit Opener & Lesson 1 What can you tell about an object's position?			
	Activity	**Pages**	**Time**
Engage	• Unit Opener: Think! *Does the ball move on its own?* • Unit Opener: Identify changes in position. • Unit Opener: Identify some ways objects move. • Think! *Can an object be above and next to another object?*	SB p. 100 SB p. 100 SB p. 100 SB p. 103	10 min 10 min 10 min 5 min
Explain	• Position: on, above, and below • Position: in front of and behind • Position: next to	SB p. 101 SB p. 102 SB p. 103	30 min 30 min 30 min
Elaborate	• From Slowest to Fastest • Put the Book on the Pencil • Above and Below • Did they change position? • Flash Lab: Position of Objects	TB p. 100 TB p. 101 TB p. 101 TB p. 102 SB p. 102	5 min 5 min 10 min 10 min 10 min
Evaluate	• *Lesson 1 Check* (ActiveTeach) • Assessment for Learning • Review (Lesson 1) • *Got it? Self Assessment* (ActiveTeach) • *Got it? Quiz* (ActiveTeach)	TB p. 111a TB p. 103 SB p. 111 TB p. 111b TB p. 111b	10 min 10 min 10 min 10 min 10 min

Lesson 2 What are some ways objects move?			
	Activity	**Pages**	**Time**
Engage	• Think! *Can you be below something you push?* • Think! *What happens to the yo-yo if you stop moving your hand?* • Think! *Can you push or pull a liquid?* • Think! *Look at pictures 1 and 2. Is it easier to push or pull the sled? Why?*	SB p. 104 TB p. 105 TB p. 106 TB p. 107	10 min 10 min 10 min 10 min
Explore	• Digital Lab: *How do objects move?* (ActiveTeach)	TB p. 104	30 min
Explain	• Pushes • Pulls • Speed	SB p. 105 SB p. 106 SB p. 107	30 min 30 min 30 min
Elaborate	• Draw It • Balloon Yo-Yos • At-Home Lab: Things We Push and Pull at Home • Fast or Slow	TB p. 105 TB p. 106 SB p. 106 TB p. 107	10 min 10 min 10 min 10 min
Evaluate	• *Lesson 2 Check* (ActiveTeach) • Assessment for Learning • Review (Lesson 2) • *Got it? Self Assessment* (ActiveTeach) • *Got it? Quiz* (ActiveTeach)	TB p. 111a TB p. 107 SB p. 111 TB p. 111b TB p. 111b	10 min 10 min 10 min 10 min 10 min

<table>
<tr><td colspan="4">**Lesson 3 What are magnets?**</td></tr>
<tr><td></td><td>**Activity**</td><td>**Pages**</td><td>**Time**</td></tr>
<tr><td>**Engage**</td><td>• Think! *Which does the magnet attract faster, the paper clip or the screw?*</td><td>TB p. 108</td><td>10 min</td></tr>
<tr><td rowspan="3">**Explain**</td><td>• How magnets attract</td><td>SB p. 108</td><td>30 min</td></tr>
<tr><td>• How magnets repel</td><td>SB p. 109</td><td>30 min</td></tr>
<tr><td>• *Got it? 60-Second Video* (ActiveTeach)</td><td>TB p. 109</td><td>10 min</td></tr>
<tr><td>**Elaborate**</td><td>• A Paper Clip Chain</td><td>TB p. 109</td><td>10 min</td></tr>
<tr><td rowspan="5">**Evaluate**</td><td>• *Lesson 3 Check* (ActiveTeach)</td><td>TB p. 111a</td><td>10 min</td></tr>
<tr><td>• Assessment for Learning</td><td>TB p. 109</td><td>10 min</td></tr>
<tr><td>• Review (Lesson 3)</td><td>SB p. 111</td><td>10 min</td></tr>
<tr><td>• *Got it? Self Assessment* (ActiveTeach)</td><td>TB p. 111b</td><td>10 min</td></tr>
<tr><td>• *Got it? Quiz* (ActiveTeach)</td><td>TB p. 111b</td><td>10 min</td></tr>
<tr><td>**Lab**</td><td>• *Let's Investigate! How can you move the car?* (ActiveTeach)</td><td>SB p. 110</td><td>30 min</td></tr>
</table>

Flash Cards

Lesson 1	
Key Words	**ELL Support**
position, on, above, below, in front of, behind, next to	**Vocabulary:** *duckling, sled, bird, purple, green, black, white*

Lesson 2	
Key Words	**ELL Support**
push, away from, toward, pull, fast, slow	**Vocabulary:** *move, swing, kick, wagon, sled, rope, yo-yo, change, train, tricycle*

Lesson 3	
Key Words	**ELL Support**
magnet, attract, repel	**Vocabulary:** *metal, paper clip, screw, (copper) pennies, building block, nail, crayon, coin*

Unit 9 — Motion

Unit Objectives

Lesson 1: Students will learn about position.

Lesson 2: Students will learn about how objects can move.

Lesson 3: Students will learn about magnets.

Vocabulary: *position, on, above, below, in front of, behind, next to, push, away from, toward, pull, fast, slow, magnet, attract, repel*

Introduce the Big Question

What are position and motion?

Build Background

Help students consolidate concepts. Invite each student to choose an item from their pencil case or backpack and place it on your table. Pick items randomly and have students identify them. Discuss their shapes, sizes, textures, etc. *What are all these items made of? Matter. What are three types of matter? Solid, liquid, gas.*

Activate prior knowledge by reminding students that scientists use their senses to learn about the world around us. *One way we can learn about things is to know about their position and motion. We're going to learn more about position and motion in this Unit.* Ask the Big Question and have students predict what they think the concepts mean. Accept all logical answers.

Engage

Think!

Does the ball move on its own?

Draw students' attention to the picture of the boys playing soccer on the lower right of the page. Read the question aloud. Allow students time to discuss. Monitor and provide support as needed.

① Look at picture 1. Look at picture 2. What changes? Talk as a class.

Point to the pictures and ask questions. *Is the ball a living thing or a nonliving thing? Nonliving! Can the ball move on its own? No! Where is the ball in picture 1? Do you think the ball moved in picture 2? Yes!* Invite students to talk about or point to the different positions of the ball in pictures 1 and 2. *How do you think the ball moved? The boys moved it.*

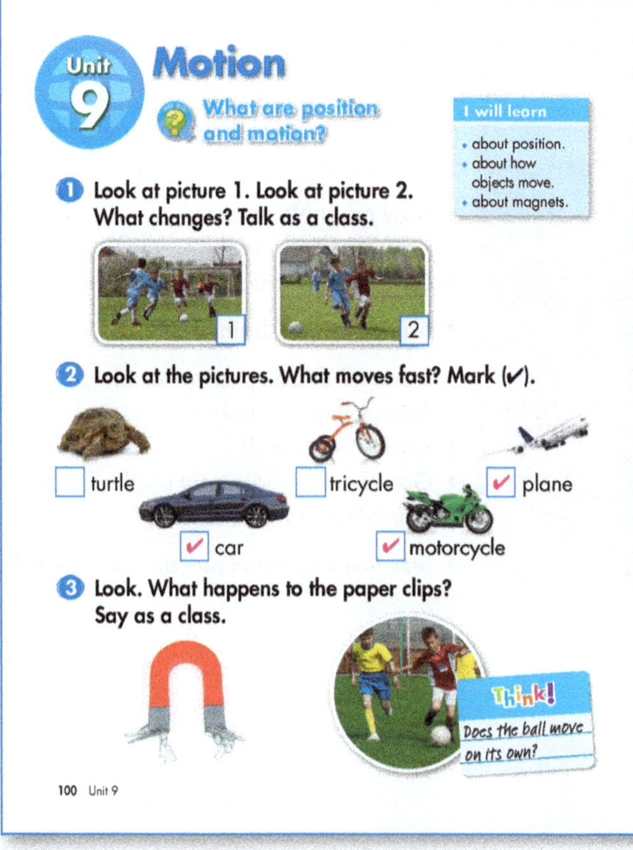

② Look at the pictures. What moves fast? Mark (✔).

Point to the pictures one by one and elicit or identify what's pictured. Read the question aloud. *Does a car get you to school fast? Yes! Does a turtle run as fast as you? No!* Check answers as a class.

③ Look. What happens to the paper clips? Say as a class.

Draw students' attention to the picture of the magnet and the paper clips. Pre-teach *magnet*. *Have you ever seen a magnet? What can you do with a magnet?* Accept all logical answers. Read the question aloud and discuss it as a class. Elicit or explain that the magnet pulls the paper clips.

Elaborate

From Slowest to Fastest

Have students number the pictures in exercise 2 from 1 (slowest) to 5 (fastest). *Can a car go as fast as a plane? Can a tricycle go as fast as a motorcycle?* Check answers as a class. Encourage students to explain their reasoning. (Possible answers: *My brother's motorcycle goes really fast. My turtle moves really slow.*)

Think! Again!

Revisit the question *Does the ball move on its own?* Invite students to share their ideas freely. (Possible answers: *No, it doesn't move on its own. The boys kick it.*) Provide vocabulary support as needed and accept all logical answers.

What can you tell about an object's position?

Objective: Learn about an object's position.

Vocabulary: *position, on, above, below*

Digital Resources: Flash Cards (*on, above/below*)

Materials: variety of classroom objects (pencil case, pens, pencils, crayons, notebook, book, glue, scissors, etc.), big piece of cloth or scarf to use as a cover

Unlock the Big Question

Write the following on the board: *I will learn about an object's position.*

Build Background

Set out different classroom objects on your table. Invite volunteers to identify the objects by saying *There's a (ruler) on the table.* Then cover the objects with the cloth or scarf. Say true/false sentences about the objects on the table. *There's a backpack on the table.* (*False. There isn't one.*) *There's a green pencil case on the table.* (*False. It's red.*) Each time, have students remember and say whether you are right or wrong.

Explain

1. **Read. Look at picture 2. Where is the ball? Say with a partner.**

   Read the paragraph aloud for students. Show the *on* and *above/below* Flash Cards and check comprehension. *Where is the ball? Where is the purple notebook? Where are the glasses?* Elicit and accept all logical answers.

   Ask students to explain what we talk about when we describe the position of an object. (Possible answers: *Where it is. We can use other objects to say where it is.*)

2. **Look around the classroom. What is above you? What is below you? Talk as a class.**

   Read the questions aloud one by one and have students look around the classroom. Invite volunteers to say what is above and below them. Provide vocabulary support as needed.

3. **Pretend you are outside. What is above you? What is below you? Talk with a partner.**

   Ask students to pretend they are outside in the school playground, park, or countryside. Pair students and have them discuss what they can

---

Lesson 1 · What can you tell about an object's position?

Key Words
- position
- on
- above
- below
- in front of
- behind
- next to

1. **Read. Look at picture 2. Where is the ball? Say with a partner.**

Above and Below

Position is where objects are. An object can be on another object. An object can be above another object. An object can be below another object. Look at picture 1. The apple is on the paper. The apple is above the books. The books are below the apple.

2. **Look around the classroom. What is above you? What is below you? Talk as a class.**

3. **Pretend you are outside. What is above you? What is below you? Talk with a partner.**

Unit 9   101

---

see above and below them. They can talk about the sky, clouds, the sun, trees, grass, soil, etc. Monitor and provide support as necessary. Invite volunteers to share their ideas with the class.

Elaborate

Put the Book on the Pencil

Have students gather the same items you used at the beginning of class in front of them, for instance, a book, a pencil case, a crayon, an eraser, a ruler, a piece of paper, etc. Give students instructions. *Put the book below the crayon. Hold the eraser above the crayon.* Monitor and make sure students are following your instructions. Invite volunteers to give similar instructions to the class.

Above and Below

Play a quick game of Simon Says. *Simon says... put your hands above your head!* Students put their hands above their heads. *Simon says... put your hands below your knees.* Students put their hands below their knees. *Put your foot above your desk.* Students stand still.

Alternatively, show students how to make a "perpetual bridge": Students make two lines, stand facing one another, and hold their arms up with their partner to make a "bridge." The pair at one end lets go, goes under the bridge, and shouts *Below!* When they reach the end, they form part of the bridge again, and shout *Above!*

What can you tell about an object's position?

> **Objective:** Learn more about the position of objects.
>
> **Vocabulary:** *in front of, behind, duckling, sled*
>
> **Digital Resources:** Flash Cards (*on, above/below, in front of/behind*), *I Will Know...* Digital Activity

Build Background

Display the *on* Flash Card and have students look at it for a minute or less. Then hide it and ask questions to check their memory. *What's on the green notebook? (A red notebook.) What's below the glasses? (A small book/notebook.) What else is on the small book/notebook? (An apple and a pencil holder with pencils.)*

Explain

4 **Read. Look at the picture of the ducklings. Draw another duckling behind duckling number 5.**

Draw students' attention to the picture of the ducklings. *How many ducklings can you see? Five. Are they big or small? Small.* Read the paragraph aloud and have students draw another duckling behind duckling number 5. Monitor and provide support.

5 **Look at the photo of the boy and his sled. Where is the boy? Circle.**

Point to the picture of the boy and the sled. *What is the boy doing? (Playing with his sled.) What's the weather like? (It's cold and snowy.)*

Then read the question aloud. *Where is the boy?* Have students circle the correct preposition to complete the sentence. Check the answer as a class.

6 **Get in a line at the front of the classroom. Who is in front of you? Who is behind you? Say.**

Invite students or a group of students to stand up and line up at the front of the class. Have them face forward so they are standing one behind the other.

Stand next to random students and ask *Who is in front of you? Who is behind you?*

If you have a large class and you are doing this in groups, invite students who are sitting down to answer. *Who is in front of (Mary)? Who is behind (Sam)?*

> **I Will Know...**
>
> Have students do the *I Will Know...* Digital Activity.

---

4 Read. Look at the picture of the ducklings. Draw another duckling behind duckling number 5.

In Front of and Behind

Look at the ducklings. Duckling number 1 is **in front of** the other ducklings. The other ducklings are **behind** duckling number 1.

5 Look at the photo of the boy and his sled. Where is the boy? Circle.

The boy is in front of / behind the sled.

6 Get in a line at the front of the classroom. Who is in front of you? Who is behind you? Say.

> **Flash Lab**
>
> Go around the classroom. Find an object that is in front of another object. Tell a partner about the position of each object.

Elaborate

Did they change position?

Have a group of five to seven students line up again at the front of the classroom. Write the students' names on the board in the order they are standing.

Have the students who are sitting down close their eyes. Change the position of two students. Then have students open their eyes and say who changed position. Provide help with language. *(Pablo) was behind (Mary). Now he's behind (Sergio).*

Switch groups and repeat the exercise until all students have had a chance to be in the line-up and to say who changed position.

⚡ Flash Lab

Position of Objects

Have students form pairs. Have them walk around the classroom and stop to look at objects. They should take turns describing the objects' positions. Monitor and provide support. Then, as a class, elicit some of the sentences students made. (Possible answers: *The (pencil) is next to the (book). The (desk) is in front of the (chair). The (chair) is behind the (desk).*)

What can you tell about an object's position?

Objective: Learn more about the position of objects.

Vocabulary: *next to, bird, purple, green, black, white*

Digital Resources: Flash Cards (*on, above/below, in front of/behind*), *Lesson 1 Check* (print out 1 per student)

Materials: crayons (a set for each student)

Build Background

Review *on, above/below,* and *in front of/behind* with the Flash Cards. Then distribute the crayons and give instructions for students to follow. *Hold the green crayon above the yellow crayon. Put the brown crayon in front of the purple crayon.* Have students hold on to the crayons as they put them in position so that they don't fall.

Explain

7 **Read. Where is the blue bird? Say with a partner.**

Focus students' attention on the picture of the birds. Point to one bird at a time and elicit its colors. Read the paragraph aloud and check comprehension by asking the question *Where is the blue bird? It's next to the white/yellow bird.* Ask similar questions and invite volunteers to answer.

ELL Vocabulary Support

Students should now be familiar with the following prepositions of place or location: *on, above, below, in front of, behind, next to.* Write them on the board and, next to each one, draw a small picture, using a box and a ball to depict position. For example, draw the ball *on* the box to depict *on*, draw the ball *under* the box at a distance to depict *below*, etc. Have students copy the prepositions and pictures in their notebooks.

8 **Where is the yellow and green bird? How many birds are next to it? Say as a class.**

Point to the yellow and green bird. Read the question aloud and invite students to answer. (Possible answers: *It's next to the purple bird. There are five birds next to it. There is one bird next to it.*)

9 **Pretend you are at home in your bed. What is next to you? Draw.**

Have students imagine they are at home in their beds. Ask them to draw a picture of objects that are next to them. Monitor and provide support.

7 Read. Where is the blue bird? Say with a partner.

Next To

The purple bird is **next to** the green and black bird. The white bird is next to the green and black bird. The green and black bird is next to the purple bird and the white bird.

8 Where is the yellow and green bird? How many birds are next to it? Say as a class.

9 Pretend you are at home in your bed. What is next to you? Draw.

Lesson 1 Check Unit 9 103

Then invite volunteers to show their drawings to the class and explain what they have drawn. Provide support as needed.

Think!

Can an object be above and next to another object?

Read the question aloud and have students discuss in small groups. Have groups share their answers with the class and explain their reasoning. Encourage them to use objects to demonstrate how it is possible. If students are having difficulty, draw students' attention to the picture of the boy and the soccer ball and ask questions. *Is the ball above the boy's arm? Is the ball next to the boy's head?*

Evaluate

Lesson 1 Check Assessment for Learning

Review the Key Words for Lesson 1 (see Student's Book page 101). Distribute the *Lesson 1 Check* and guide students as they complete it. Check answers as a class. Then ask students to grade their progress on the topic of an object's position from 1 to 3: 3 = *I understand what position is;* 2 = *I need to study more;* 1 = *I need help!* Encourage students giving themselves a 1 or a 2 to say what they found difficult and what they need to study more.

What are some ways objects move?

Objective: Learn about some ways in which objects move.

Vocabulary: *push, away from, toward, move, swing, kick*

Digital Resources: Flash Cards (*push, pull*), *Let's Explore!* Digital Lab

Materials: a big cardboard box with a string attached to it

Unlock the Big Question

Write the following on the board: *I will learn about how objects move. We have been talking about position, or where objects are. In this lesson, we will talk about motion, or some ways objects can move.*

Build Background

Place the box in the middle of the class. *What is it? A box. Is there anything in it?* Invite students to look inside the box. Then start pushing the box across the classroom. *What am I doing?* Have students give their opinion. Then start pulling the box. *What am I doing?* Invite students to answer. *You are moving the box.*

Explore

Let's Explore! Lab   How do objects move?

Objective: Learn some ways in which objects move.

Digital Resources: *Let's Explore!* Digital Lab, *Let's Explore! Activity Card* (*Optional:* Show the Digital Lab instead.)

Materials: rolling toy car, paper, pencil

- Demonstrate how to do the activity. Place the toy car on the sheet of paper and push it gently. Trace a line to show the distance and direction the toy car moved. Repeat and trace a new line.
- Distribute the materials and have students perform the lab. Monitor and provide support.
- Have students work in pairs to complete the *Activity Card.*

Explain

1 Read. Think of three things you push. Say with a partner.

Read the paragraph aloud. Focus students' attention on the pictures of the children playing with the toy car and the boy on the swing. Put

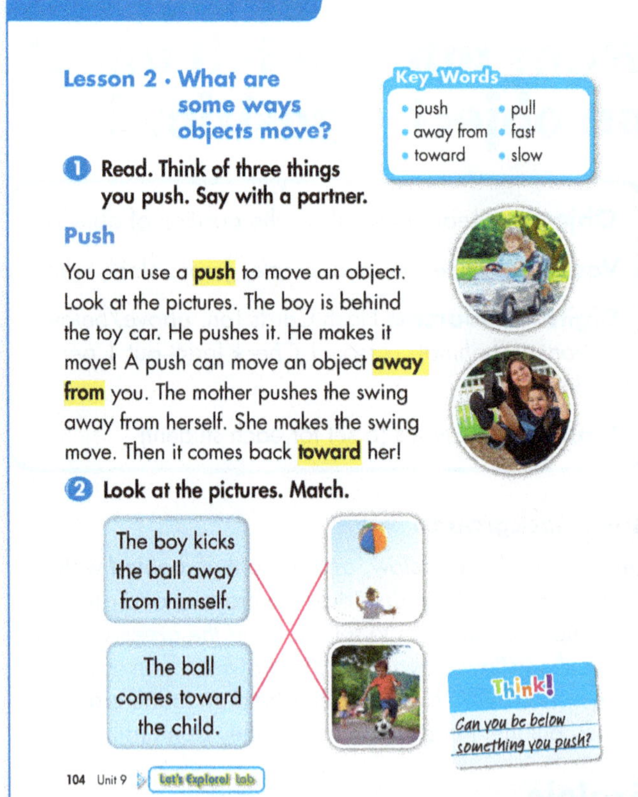

students in pairs and have them brainstorm three things they can push. Invite volunteers to share answers with the class. (Possible answers: *I can push a box. I can push a swing. I can push a chair.*) You may wish to have students demonstrate by pushing a chair or table across the floor. *You make the (chair) move! You push the (chair)!*

2 Look at the pictures. Match.

Point to the two pictures of the boys playing with a ball. *What's the same in these two pictures? The boys are playing with a ball. What are the differences?* Elicit ideas. *The boy with the beach ball is below the ball. The boy with the soccer ball is behind the ball.*

Read the two sentences aloud and have students match them to the pictures. Check answers as a class. *Right! The boy with the soccer ball pushes, or kicks, the ball away from himself. The boy with the beach ball pushes, or throws, the ball up in the air. Now it is coming back down toward him!*

Think!

Can you be below something you push?

Read the question aloud. Draw students' attention to the picture of the boy with the beach ball. Have students discuss in small groups. Have groups share their answers with the class and explain their reasoning. (Possible answers: *Yes! The boy can push a ball into the air. A throw is a push! He is below it.*)

What are some ways objects move?

Objective: Learn more about some ways in which objects can move.

Vocabulary: *a pull, toward, wagon, in front of, sled, rope, yo-yo*

Digital Resources: Flash Cards (*push, pull*), *I Will Know...* Digital Activity

Materials: yo-yo

Build Background

Display the *push* and *pull* Flash Cards. Elicit what movements students can see in each Flash Card. Have students place four classroom objects on their tables, e.g., a pencil, an eraser, a book, and a ruler. Give instructions for students to follow. *Push the ruler away from you. Pull the eraser toward you.*

Explain

3 **Read. What does a pull do? Say with a partner.**

Read the paragraph aloud for students. As you read, point to the pictures that are mentioned in the paragraph and have volunteers describe what is shown. (*They pull a wagon. She pulls a sled. They pull a rope.*) Then ask what happens when you pull something. *It moves!* Ensure students understand that, sometimes, something you pull moves toward you.

Finally, elicit some ways students use pulls every day, e.g., when they pull drawers, doors, curtains, etc.

4 **Look at pictures 1–4. Mark (✔) the pictures that show a pull.**

Have students mark the pictures that show a pull. Check answers as a class and discuss what the pull is used for in each picture. (Answers: *The boy pulls his backpack up. The girl pushes the books.*)

Then ask what happens in the other pictures. Provide vocabulary support as needed. (Answers: *The man pushes a shopping cart. The horse pulls a wagon.*)

5 **Look at the picture. Does the girl push or pull the yo-yo? Talk with a partner.**

Play with the yo-yo and invite students to try as well. Then draw their attention to the picture of the girl on the lower right of the page. Read the question aloud and have students discuss it in pairs. Check answers as a class. (Answer: *She pushes and pulls the yo-yo.*)

Ask questions to consolidate comprehension. *Does she push the yo-yo toward her? No! She pulls the yo-yo toward her! Does she pull it away from her? No! She pushes it away from her!*

Think!

What happens to the yo-yo if you stop moving your hand?

Ask the question and play with the yo-yo for a few seconds. Then catch it abruptly and ask the question again. Invite students to share ideas. Play with the yo-yo once more and stop moving your hand. Discuss what students see. (Answer: *The yo-yo will slow down, stop moving, and hang from the string.*)

Elaborate

Draw It

Have students draw a picture of themselves pushing something away from them and a picture of themselves pulling something toward them. They can refer to the pictures in their books if necessary. Monitor and provide support.

Then divide students into small groups and have them show their pictures and explain what they have drawn.

I Will Know...

Have students do the *I Will Know...* Digital Activity.

Lesson 2
What are some ways objects move?

> **Objective:** Learn more about pushes and pulls.
>
> **Vocabulary:** *a push, a pull, position*
>
> **Digital Resources:** Flash Cards (*push, pull*)
>
> **Materials:** per student: balloon, piece of string

Build Background

Invite a student to stand at the front of the class and give the student the Flash Cards. Push or pull an object and have the student hold up the correct Flash Card. Repeat a few times with different students and pushing or pulling different objects.

Explain

6 Look at the pictures that show pushes and pulls. Match.

Have students look at the pictures and match them to *push* or *pull*. Invite them to compare with a partner. Check answers as a class.

Have volunteers describe a picture and explain what is being pushed or pulled and what happens. (Possible answers: *The family pushes bicycles. The bicycles are next to them. The parents pull the child up. She swings.*) Provide vocabulary support as needed.

7 How are a push and a pull alike? Talk as a class.

Read the question aloud and elicit ideas from the class. Discuss how both pushes and pulls are ways we can move things. *What can happen to an object when you push it? It can move! What can happen to an object when you pull it? It can move! Right! Both pushes and pulls can move an object.*

8 You push or pull an object. What happens to its position? Talk as a class.

Read the question aloud and have students brainstorm ideas as a class. Elicit or explain that, when we push or pull an object, its position can change. Discuss how it might not change if, for example, the object is too heavy. You may want to demonstrate how you can push the classroom wall, but that you can't make it move or change its position. *I am not strong enough to make the wall move!*

Elaborate

Balloon Yo-Yos

Before class, blow up the balloons and attach a piece of string to each. Distribute the balloons and demonstrate

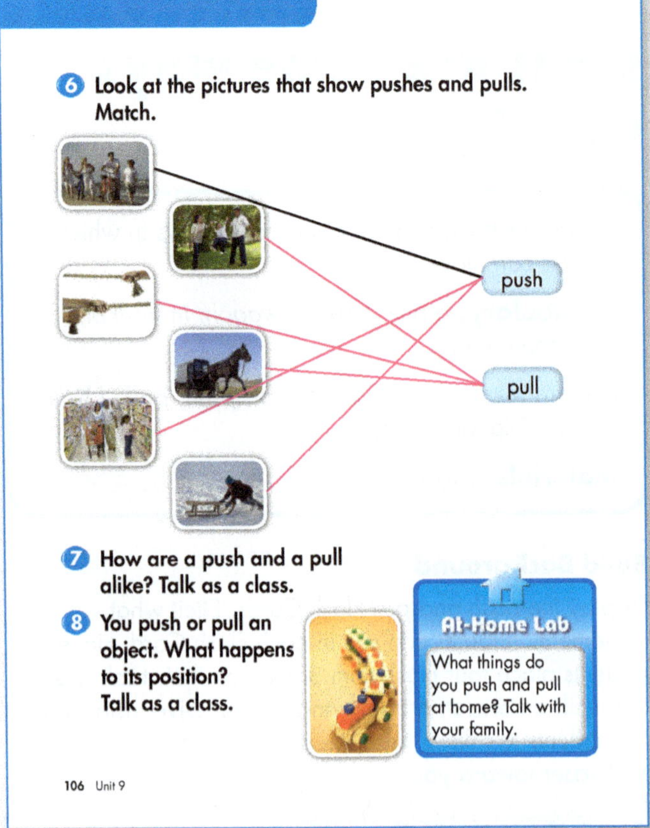

how to use them as yo-yos. Discuss with the class whether the balloon moves in the same way as a real yo-yo. (Possible answer: *No, it's lighter. It takes longer for the balloon to come back.*) *What kind of matter is a regular yo-yo? A solid! What kind of matter is a balloon? A solid that is filled with gas! Do you think that makes the balloon move differently?* Accept all logical answers.

Invite students to play with the balloons. You may wish to take students outside to play with their balloon yo-yos.

At-Home Lab

Things We Push and Pull at Home

Assign the At-Home Lab as homework. The next day, invite students to say or mime what objects they push and pull at home. Provide vocabulary support as needed.

Think!

Can you push or pull a liquid?

Read the question aloud and invite students to discuss freely. If students are having difficulty, ask them if they have ever been swimming or played with toys in a bathtub. *Can you make the water move? Can you push it away from you and pull it toward you? Yes!*

Lesson 2

What are some ways objects move?

Objective: Learn about objects you can make move fast or slow.

Vocabulary: *fast, slow, push, pull, position, change, move, train, tricycle*

Digital Resources: Flash Cards (*fast, slow*), *Lesson 2 Check* (print out 1 per student)

Materials: variety of Animal Cards (*camel, coyote, crocodile, deer, dolphin, elephant, horse, kangaroo, snail, crab, turtle, bear,* etc.; 4 per group)

Build Background

Display the Flash Cards and review or pre-teach *fast* and *slow*. Divide the class into small groups. Give students a minute to brainstorm as many fast and slow objects or animals as they can. Elicit ideas from the class. Discuss which of the objects or animals mentioned is the fastest and slowest.

Explain

9 Read. What are some things you can make move fast? Talk as a class.

Read the paragraph aloud for students. Draw students' attention to the photos. Read the captions and the question aloud. You may want to refer to other photos in the Unit as well. Elicit ideas from students. (Possible answers: *I can make my toy car move fast. I can make a soccer ball move fast.*)

10 Do these things move fast or slow? Say with a partner.

Draw students' attention to the four pictures and invite them to say what they see. Put students in pairs and have them say whether the thing in each picture moves fast or slow. Check answers as a class. (fast: *race car, plane;* slow: *turtle or tortoise, baby*)

Think!

Look at pictures 1 and 2. Is it easier to push or pull the sled? Why?

Read the question aloud and allow students to discuss freely. Accept all logical answers. (Note: It takes the same amount of force to push or pull the same object in the same circumstances. In the instances pictured, it might be easier to pull the sled.)

---

9 Read. What are some things you can make move fast? Talk as a class.

Fast and Slow

A push or pull can change the position of an object. A push or pull can change how an object moves. A push or pull can make an object move **fast** or **slow**.

The train moves fast.

The tricycle moves slow.

10 Do these things move fast or slow? Say with a partner.

1

2

Lesson 2 Check Unit 9 107

Elaborate

Fast or Slow

Show the Animal Cards one by one and have students identify the animals. Divide the class into small groups and distribute the cards. Each group should have at least four cards. Invite students to put the cards in order, with the slowest animal on the left and the fastest one on the right. Monitor and provide support. Accept all logical answers.

Then have students walk around the tables and look at the cards. Invite students to say whether they agree or disagree with the order and why.

ELL Vocabulary Support

In American English, both *slow* and *slowly* can be used as adverbs. *Slow* is more typically used in phrases in which *slow* follows the verb, for example, *Go slow!* In spoken language, *slow* is often used, whereas, in written English, *slowly* is preferred.

Evaluate

Lesson 2 Check Assessment for Learning

Review the Key Words for Lesson 2 (see Student's Book page 104). Distribute the *Lesson 2 Check* and guide students as they complete it. Check answers as a class. Then ask students to grade their progress on the topic of how objects can move from 1 to 3: 3 = *I understand how objects can move;* 2 = *I need to study more;* 1 = *I need help!* Encourage students giving themselves a 1 or a 2 to say what they found difficult and what they need to study more.

What are magnets?

Objective: Learn about how magnets can move some objects.

Vocabulary: *magnet, attract, metal, pull toward, paper clip, screw, pennies*

Digital Resources: Flash Card (*magnet*), *I Will Know...* Digital Activity

Materials: a big magnet, a paper clip, a screw, an eraser, a crayon, smaller magnets (1 per student or pair of students), copper coins, wire, or jewelry

Unlock the Big Question

Write the following on the board: *I will learn how magnets can move some objects.*

Build Background

Pre-teach *magnet* by showing the Flash Card. Invite students to say what they know about magnets and whether they have ever used one. Then display the paper clip, the screw, the eraser, and the crayon on your table. Have students say which objects they think the magnet will move, but don't check answers yet. *In this lesson, we will learn about some special kinds of pushes and pulls.*

Think!

Which does the magnet attract faster, the paper clip or the screw?

Ask the question and have students discuss freely. Then have students place the paper clip and the screw side by side. Have them slide the magnet on the table slowly, bringing it closer to the paper clip and the screw. Invite them to observe which object will move toward the magnet first. If necessary, do this example as a class. Discuss why the magnet did or didn't move one object faster.

Explain

①　Read. What does attract mean? Say with a partner.

Read the paragraph aloud for students. *What does attract mean?* (*It means to pull toward.*) Invite students to take the big magnet and place it close to one of the objects on your table. Elicit whether the magnet attracts the object and whether it makes the object move. (Possible answers: *The magnet makes the paper clip move toward it. It doesn't make the crayon move.*)

Lesson 3 · What are magnets?

①　Read. What does attract mean? Say with a partner.

Key Words
- magnet
- attract
- repel

Magnets Attract

Magnets attract some metal objects. Attract means to pull toward. Magnets attract paper clips and screws. Magnets do not attract objects that are not made of metal, like crayons. They do not attract some metal objects, like pennies.

Magnets attract paper clips and screws.

Magnets do not attract erasers, pencils, or copper pennies.

②　What objects do magnets attract? What objects do magnets not attract? Talk as a class.

③　Look around the classroom. Find two things a magnet attracts and three things it does not attract.

108　Unit 9　*I Will Know...*

②　What objects do magnets attract? What objects do magnets not attract? Talk as a class.

Ask the questions and have students look at the pictures in their books. (Possible answers: *Magnets attract paper clips and screws. They are made of metal. Magnets don't attract crayons or pencils. They are not made of metal.*) Explain that magnets don't attract pennies because pennies are made of copper. *Magnets don't attract things made of copper.*

③　Look around the classroom. Find two things a magnet attracts and three things it does not attract.

Have students look around the classroom to find the five objects. You may have to put some things the magnets will attract around the classroom ahead of time. If possible, give students a magnet and have them walk around the classroom to test their ideas. Elicit ideas from the class. (Possible answers: *It attracts the stapler and scissors. It doesn't attract my backpack, desktop, or window.*)

You may want to explain that magnets can attract some objects but can't always make them move. *The magnet attracts the (desk leg). The magnet sticks to the (desk leg). But, it can't make it move. The (desk leg) is attached to the (desk), so the magnet can't make it move.*

I Will Know...

Have students do the *I Will Know...* Digital Activity.

What are magnets?

Objective: Learn more about magnets.

Vocabulary: *magnet, attract, repel, push away, paper clip, building block, nail, crayon, screw, copper coin*

Digital Resources: Flash Card (*magnet*), *Lesson 3 Check* (print out 1 per student), *Got it? 60-Second Video*

Materials: 1 of each per pair or small group: magnet, paper clip, building block, nail, crayon, screw, copper coin (If copper coins aren't available, substitute copper wire or jewelry.)

Build Background

Hold up the Flash Card and say true/false sentences about what objects magnets attract. For example, *A magnet can attract a metal spoon.* (*Yes.*) *A magnet can attract a book.* (*No.*) *A magnet can attract a plastic plate.* (*No.*) Each time have students shout *Yes* or *No*.

Explain

4 **Mark (✔) the objects magnets attract.**

Draw students' attention to the pictures in the table. Divide the class into pairs or small groups and distribute the materials. Have students try the magnet with each object and mark the pictures in the table. Monitor and provide support. Then check answers as a class. Encourage students to support their answers. (*The building block is not made of metal.*)

5 **Read. What does repel mean? Say with a partner.**

Read the paragraph aloud and invite students to explain what *repel* means. (*Push away.*) Draw their attention to the illustrations and read the captions aloud. Explain that *N* represents north and *S* represents south. Point out that the second illustration could also show south-south.

Have students use their magnets to see how two magnets can attract or repel each other. Point out how the magnets can make one another move when the like poles are end to end. Let students feel the force, if possible, or hold the magnets together and then let go so students can see the magnets move.

6 **Draw a magnet. Draw arrows to show that it attracts something.**

Read the instructions aloud and, if necessary, have a volunteer draw a picture on the board. Monitor and provide support as needed.

---

4 Mark (✔) the objects magnets attract.

Object	Attract	Don't Attract
paper clip	✔	
building block		✔
nail	✔	
crayon		✔
screw	✔	
copper coin		✔

5 **Read. What does repel mean? Say with a partner.**

Magnets Repel

Sometimes a magnet attracts another magnet. Sometimes a magnet repels another magnet. Repel means to push away.

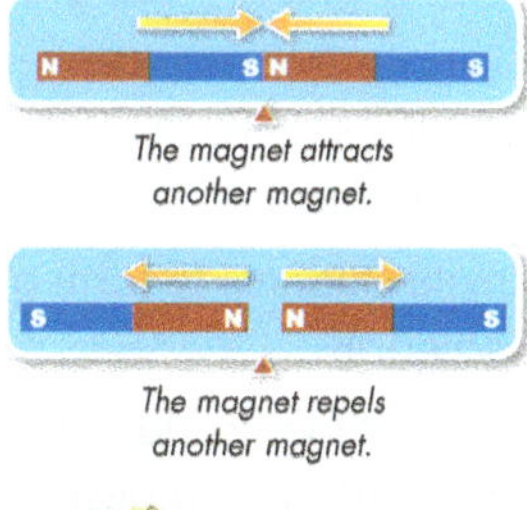
The magnet attracts another magnet.

6 **Draw a magnet. Draw arrows to show that it attracts something.**

The magnet repels another magnet.

Lesson 3 Check | Got it? 60-Second Video | Unit 9 109

Elaborate

A Paper Clip Chain

Take a magnet and stick one paper clip to it. (If the magnet is so strong that it attracts the paper clip on its side, have a student hold the paper clip straight.) Then place another paper clip at the end of the first paper clip. Continue with as many paper clips as possible. Invite students to try this in pairs or groups.

Discuss what is happening. *Why do the paper clips stick to one another?* Explain that, when the first clip touches the magnet, it becomes a temporary magnet itself, and so on. As you add clips, the temporary magnets become weaker.

Evaluate

Lesson 3 Check   Assessment for Learning

Review the Key Words for Lesson 3 (see Student's Book page 108). Distribute the *Lesson 3 Check* and guide students as they complete it. Check answers as a class. Then ask students to grade their progress on the topic of magnets from 1 to 3: 3 = *I understand how magnets can attract and repel some objects;* 2 = *I need to study more;* 1 = *I need help!* Encourage students giving themselves a 1 or a 2 to say what they found difficult and what they need to study more.

Got it?   60-Second Video

Play the *Got it? 60-Second Video* to review the unit material.

Let's Investigate!

In this unit, students learn about the position of objects and how they can move. In this lab, they will investigate how they can move a toy car using different kinds and amounts of force.

Let's Investigate! Lab How can you move the car?

Objective: Students will observe ways they can move a toy car.

Materials: per group: toy car with string attached

Digital Resources: *Let's Investigate!* Digital Lab, *Let's Investigate! Activity Card* (1 per student)

Advance Preparation: Move furniture in the classroom so that students have space to perform the experiment.

- Invite students to form groups, and distribute materials.
- Have students place their cars on the floor. Invite students, first, to push the toy car, and, then, to push the car with more force.
- Invite students to discuss what they observe.
- Then invite students, first, to pull the toy car using the string, and, then, to pull the car with more force.
- Invite students to discuss what they observe.
- Have students complete the *Activity Card*.
- You may wish to ask questions to reinforce understanding. *Does a push move the car toward you?* (No. A push moves the car away from me.) *Does a hard push move the car more?* (Yes!)

Teacher Time-Saving Option: Show the *Let's Investigate!* Digital Lab as an alternative to the hands-on lab activity.

Unlock the Big Question

Have students refer to the Big Question on the Unit Opener page. In pairs have them recall what they have learned about the position and motion of objects. Have pairs complete questions 4 and 5 on the *Activity Card*.

Let's Investigate!

How can you move the car?

1. Push.
   Push hard.
   Tell.
2. Pull.
   Pull hard.
   Tell.
3. Draw.

Class Project: Motion Poster

Materials: magazine pictures, construction paper, glue

Find images (in magazines or on the Internet) of objects or animals that are moving. Cut them out or print them out and bring them to class.

Display the pictures on a table and have students look at them. Talk about each picture as a class, discussing how fast the object or animal is moving, the direction it is moving (away, toward), and what's making it move (a push or pull). You may need to explain to students that some animals use their body parts to both push and pull themselves, for example, when they run.

Then invite students to help you create a Motion Poster. Have them glue the pictures on the paper and then invite them to draw arrows or lines to show the direction each object or animal is moving. You may divide students into groups to create a poster each. Display the poster(s) around the classroom.

Unit 9 Review

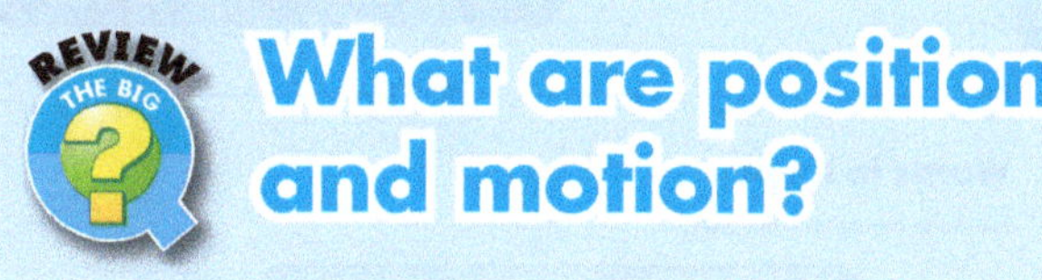

REVIEW THE BIG ? — What are position and motion?

Digital Resources: Print out 1 of each per student: *Got it? Self Assessment, Got it? Quiz*

Evaluate

Strategies for Targeted Review

The following are strategies for providing targeted review for students if they encounter challenges with the content.

Lesson 1 What can you tell about an object's position?

Question 1

If… students are having difficulty completing the sentence, then… review the prepositions of position as a class.

Lesson 2 What are some ways objects move?

Question 2

If… students are having difficulty choosing the correct word, then… ask them to look back at Lesson 2.

Lesson 3 What are magnets?

Question 3

If… students are having difficulty completing the definitions, then… demonstrate what *attract* and *repel* mean by moving your hands closer together and then farther apart.

ELL Language Support

Before students start working on the Review activities, have them read each question aloud along with you.

Got it? Self Assessment

Immediately after students have completed the Review activities, distribute a *Got it? Self Assessment* to each student. Have students complete the *Stop! Wait!* and *Go!* statements for each lesson, allowing them to look back through the lesson material if necessary.

Got it? Quiz

Distribute a Unit 9 *Got It? Quiz* to each student. Quizzes may be used for assessing students' understanding of unit concepts as well as for grading purposes.

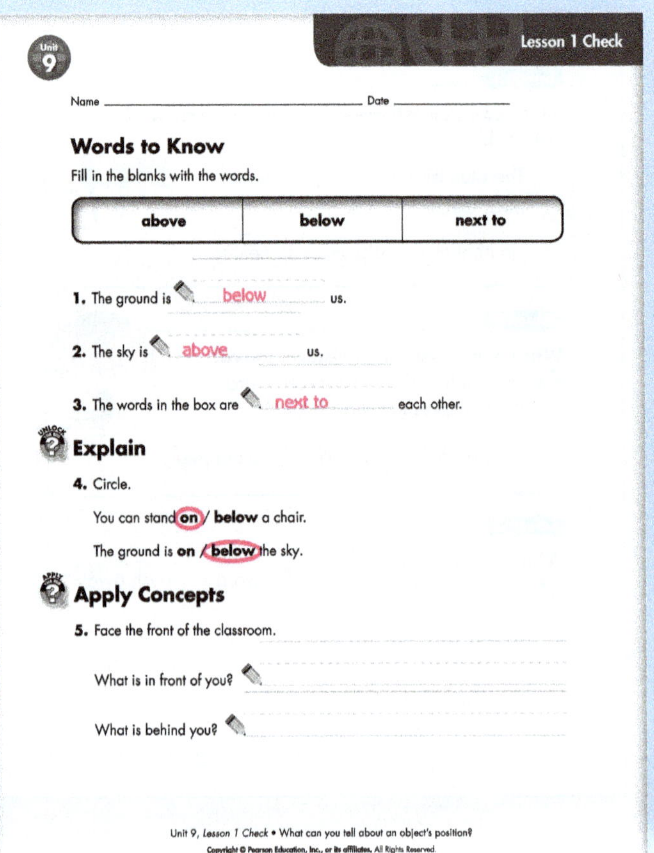

Words to Know

Fill in the blanks with the words.

above	below	next to

1. The ground is ✎ _below_ us.

2. The sky is ✎ _above_ us.

3. The words in the box are ✎ _next to_ each other.

Explain

4. Circle.

You can stand (on) / **below** a chair.

The ground is **on** / (below) the sky.

Apply Concepts

5. Face the front of the classroom.

What is in front of you? ✎

What is behind you? ✎

Unit 9, Lesson 1 Check • What can you tell about an object's position?
Copyright © Pearson Education, Inc., or its affiliates. All Rights Reserved.

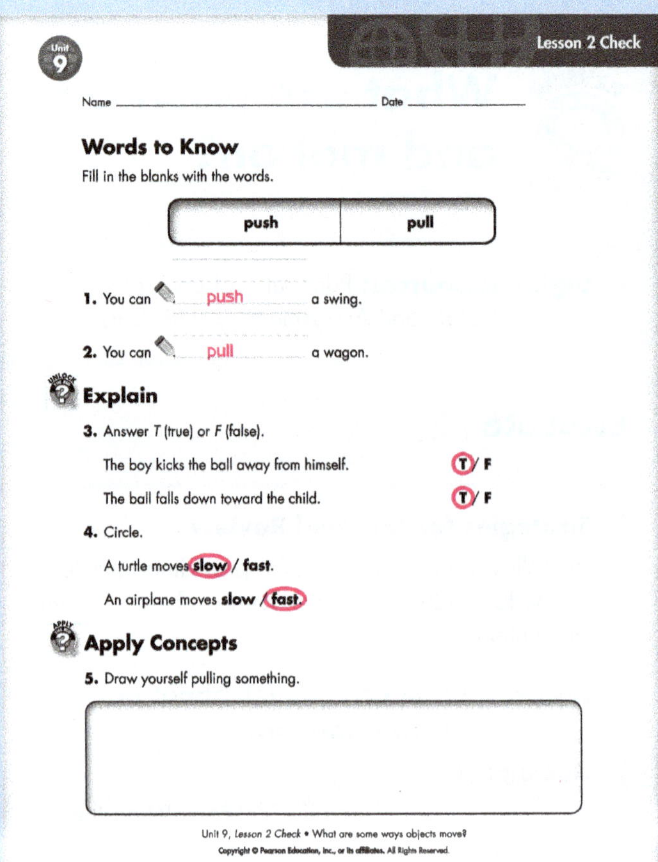

Words to Know

Fill in the blanks with the words.

push	pull

1. You can ✎ _push_ a swing.

2. You can ✎ _pull_ a wagon.

Explain

3. Answer *T* (true) or *F* (false).

The boy kicks the ball away from himself. (T) F

The ball falls down toward the child. (T) F

4. Circle.

A turtle moves (slow) / **fast**.

An airplane moves **slow** / (fast).

Apply Concepts

5. Draw yourself pulling something.

Unit 9, Lesson 2 Check • What are some ways objects move?
Copyright © Pearson Education, Inc., or its affiliates. All Rights Reserved.

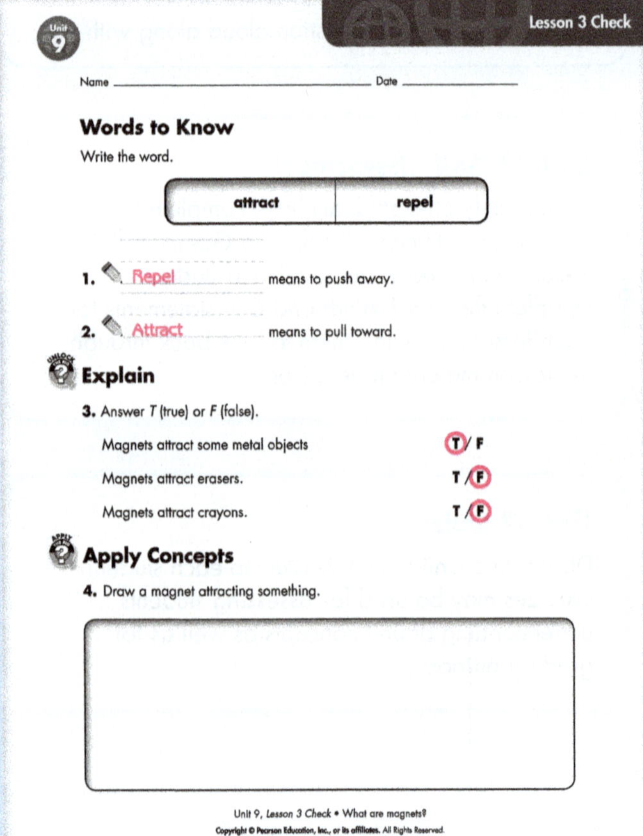

Words to Know

Write the word.

attract	repel

1. ✎ _Repel_ means to push away.

2. ✎ _Attract_ means to pull toward.

Explain

3. Answer *T* (true) or *F* (false).

Magnets attract some metal objects. (T) F

Magnets attract erasers. T (F)

Magnets attract crayons. T (F)

Apply Concepts

4. Draw a magnet attracting something.

Unit 9, Lesson 3 Check • What are magnets?
Copyright © Pearson Education, Inc., or its affiliates. All Rights Reserved.

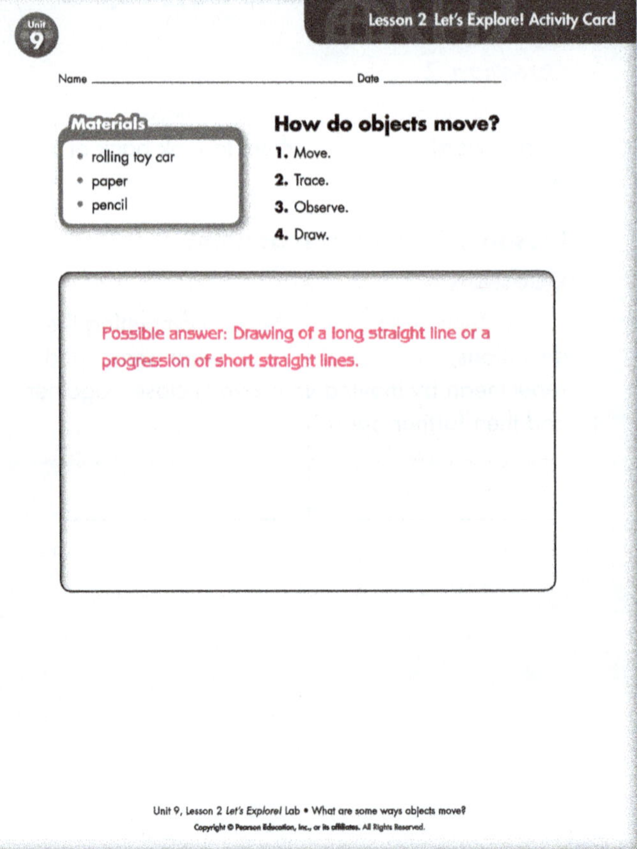

Materials
- rolling toy car
- paper
- pencil

How do objects move?

1. Move.
2. Trace.
3. Observe.
4. Draw.

Possible answer: Drawing of a long straight line or a progression of short straight lines.

Unit 9, Lesson 2 Let's Explore! Lab • What are some ways objects move?
Copyright © Pearson Education, Inc., or its affiliates. All Rights Reserved.

Name _______________________ **Date** _______________

How can you move the car?

4. Record the pushes.

Push	Push Hard
Drawing of a hand pushing the car and it going a short distance.	Drawing of a hand pushing the car and it going a longer distance.

5. Record the pulls.

Pull	Pull Hard
Drawing of a hand pulling the car and it going a short distance.	Drawing of a hand pulling the car and it going a longer distance.

Name _______________________ **Date** _______________

Got it? Self Assessment

Complete the statements for each lesson.

Lesson 1 What can you tell about an object's position?

Stop! I need help with _______________________

____

Wait! I have a question about _______________________

Go! Now I know _______________________

Lesson 2 What are some ways objects move?

Stop! I need help with _______________________

Wait! I have a question about _______________________

Go! Now I know _______________________

____

Lesson 3 What are magnets?

Stop! I need help with _______________________

Wait! I have a question about _______________________

____

Go! Now I know _______________________

Name _______________________ **Date** _______________

Got it? Quiz

Circle the best answers.

1. Position is _____________. Circle.

A all around us
B where objects are
C above us

2. The sun is _____________ us. Circle.

A above
B below
C away from

3. The ground is _____________. Circle.

A above
B below
C away from

4. One car is _________ the other. The other car is _________ the first. Circle.

A in front of / below
B next to / above
C in front of / behind

Name _______________________ **Date** _______________

5. A _____________ can move an object _____________ you. Circle.

A push / next to
B push / away from
C push / pull

6. When you run, you go _____________. Circle.

A slow
B fast
C above

7. When you push an object, you change its _____________. Circle.

A size
B color
C position

8. A magnet does not attract _____________. Circle.

A a metal paper clip
B a pencil
C a nail

Unit 9 Study Guide

What are position and motion?

Lesson 1
What can you tell about an object's position?

- Position is where an object is.
- Some ways to describe an object's position are *above*, *below*, *on*, *next to*, *in front of*, and *behind*.

Lesson 2
What are some ways objects move?

- You can push an object.
- You can pull an object.
- Objects can move away from you or toward you. Objects can move fast and slow.

Lesson 3
What are magnets?

- Magnets attract some objects, repel others, and do not affect others.

Review the Big Question

What are position and motion?

Encourage students to answer the following question in their own words:

How has your answer to the Big Question changed since the beginning of the unit? What are some things you learned that caused your answer to change?

Make a Concept Map

Draw on the board a concept map like the one shown on this page. With the students, talk through the key ideas from this unit. Invite different students to point to the ideas on the board, miming as possible.

Unit 9 Concept Map

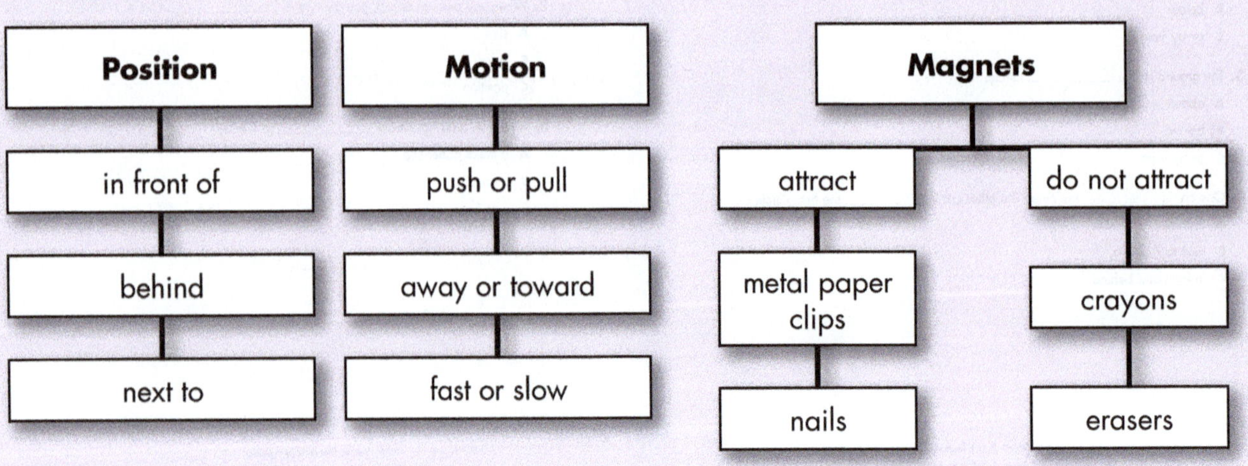

Students can make a concept map to help review the Big Question.